Measuring Safety Management Performance

Measuring Safety Management Performance lists and explains the difference between lagging and leading measures of safety management performance. It informs the reader how to use both proactive and reactive safety performance indicators and explains that consequence measurement is not an accurate reflection of the organization's safety effort. It suggests managements' Safety Performance Indicators (SPI) should be changed to proactive, positive measures of action and activities which can be controlled and accurately measured. A roadmap of a holistic system for measurement is offered that covers health and safety performance. It shows how management is traditionally informed about where they have been by information provided relating to injury data, rather than proactive, measurable, and controllable data on accident prevention efforts provided by the health and safety management system (SMS), which indicate where they are going.

This highly practical book features examples of safety performance indicators, provides positive guidelines for accurate safety performance measurement, and is based on actual workplace experiences. It explains the strengths and weaknesses of proactive and reactive measurement metrics and gives examples of leading and lagging safety performance indicators.

This book will be an ideal read for professionals and graduate students in the fields of occupational health and safety, ergonomics, and human factors engineering. It will have resonance with managers and professionals engaged in health and safety provisions at their place of work.

Ron C. McKinnon is an internationally experienced and acknowledged safety professional, author, motivator, and presenter. He has been extensively involved in safety research concerning the cause, effect, and control of accidental loss, near miss incident reporting, accident investigation, and safety promotion. He specializes in safety culture change and the implementation and auditing of safety management systems.

Workplace Safety, Risk Management, and Industrial Hygiene
Series Editor: Ron C. McKinnon, McKinnon & Associates, Western Cape, South Africa

This new series will provide the reader with a comprehensive selection of publications covering all topics appertaining to the health and safety of people at work, at home, and during recreation. It will deal with the technical aspects of the profession, as well as the psychological ramifications of safe human behavior. This new series will include books covering the areas of accident prevention, loss control, physical risk management, safety management systems, occupational safety, industrial hygiene, occupational medicine, public safety, home safety, recreation safety, safety management, injury management, near miss incident management, school safety guidance, and other related areas within the Occupational Health and Safety discipline.

Safety Management: Near Miss Identification, Recognition, and Investigation
Ron C McKinnon

Changing the Workplace Safety Culture
Ron C McKinnon

Risk-based, Management-led, Audit-driven, Safety Management Systems
Ron C McKinnon

The Design, Implementation, and Audit of Occupational Health and Safety Management Systems
Ron C McKinnon

Cause, Effect, and Control of Accidental Loss with Accident Investigation Kit
Ron C McKinnon

A Practical Guide to Effective Workplace Accident Investigation
Ron C McKinnon

The Cause, Effect and Control of Accidental Loss, *Second Edition*
Ron C McKinnon

For more information on this series, please visit: https://www.crcpress.com/Workplace-Safety-Risk-Management-and-Industrial-Hygiene/book-series/CRCWRPLSAFRK

Measuring Safety Management Performance

Ron C. McKinnon

CRC Press
Taylor & Francis Group
Boca Raton London New York

CRC Press is an imprint of the
Taylor & Francis Group, an **informa** business

Designed cover image: © Shutterstock

First edition published 2024
by CRC Press
2385 NW Executive Center Drive, Suite 320, Boca Raton FL 33431

and by CRC Press
4 Park Square, Milton Park, Abingdon, Oxon, OX14 4RN

CRC Press is an imprint of Taylor & Francis Group, LLC

© 2024 Ron C. McKinnon

Reasonable efforts have been made to publish reliable data and information, but the author and publisher cannot assume responsibility for the validity of all materials or the consequences of their use. The authors and publishers have attempted to trace the copyright holders of all material reproduced in this publication and apologize to copyright holders if permission to publish in this form has not been obtained. If any copyright material has not been acknowledged please write and let us know so we may rectify in any future reprint.

Except as permitted under U.S. Copyright Law, no part of this book may be reprinted, reproduced, transmitted, or utilized in any form by any electronic, mechanical, or other means, now known or hereafter invented, including photocopying, microfilming, and recording, or in any information storage or retrieval system, without written permission from the publishers.

For permission to photocopy or use material electronically from this work, access www.copyright.com or contact the Copyright Clearance Center, Inc. (CCC), 222 Rosewood Drive, Danvers, MA 01923, 978-750-8400. For works that are not available on CCC please contact mpkbookspermissions@tandf.co.uk

Trademark Notice: Product or corporate names may be trademarks or registered trademarks and are used only for identification and explanation without intent to infringe.

ISBN: 978-1-032-41090-6 (hbk)
ISBN: 978-1-032-41323-5 (pbk)
ISBN: 978-1-003-35751-3 (ebk)

DOI: 10.1201/9781003357513

Typeset in Times LT Std
by KnowledgeWorks Global Ltd.

Contents

About the Author xxi
Preface xxiii

PART I Introduction to Safety Management

Chapter 1 Safety Management 3

Introduction 3
Consequence-Based Measurements 3
Not an Accurate Indicator 3
Traditional Measurement 4
False Sense of Security 4
Safety Performance 4
No Control 5
Shifting the Paradigm 5
Conclusion 5
Summary 5

Chapter 2 Workplace Health and Safety 7

Workplace Injury and Death Statistics 7
Safety 8
    Safety Is a State 8
A Safety Measure 8
Occupational Hygiene 8
    Objective 8
    Occupational Hygiene Stresses 9
Hazard 9
Potential Hazard 9
Risk 9
Hazard Identification and Risk Assessment (HIRA) 10
Risk Assessment 10
    Components 10
Hierarchy of Control 11
Business Interruption 11
Accident 11
    Other Definitions 12
Near Miss Incident 12
Accident/Injury/Incident Confusion 13
    Clarity 13
Injury, Disease, or Illness 13

High-Risk Behavior (Unsafe Act) 14
High-Risk Condition (Unsafe Condition) 14
    Immediate Accident Causes? 14
Accident Sequence or Loss Causation Sequence (Domino Sequence) 14
Accident Causation Theories 14
    Loss and No Loss 14
A Basic Loss Causation Sequence 15
    Shift in Time or Position 15
    Narrowly Avoided 16
H.W. Heinrich 16
Losses 16
Types of Losses 16
Direct Losses (Insured Losses) 17
    Medical 17
    Compensation 17
    Permanent Disability 17
    Rehabilitation 17
Health and Safety Management System (SMS) 18
    A Continuous Process 18
Risk-based, Management-Led, and Audit-driven 18
    Risk-based 18
    Management-led 18
    Audit-driven 19
Management Safety Functions 19
    Safety Controlling 19
    Risk-based, Management-led, Audit-driven SMS 20
    Authority, Responsibility, and Accountability 22
Summary 23

Chapter 3 Understanding and Analyzing Accidental Loss Causation 25

A Sequence of Events 25
    Other Accident Causation Theories 25
Loss Causation Sequence 25
Losses 26
    Importance 26
Near Miss Incident 26
    Traditional Viewpoint 26
Failure to Identify Hazards and Assess the Risk 27
Weak or Non-Existent Safety Management System 27
Accident Root Causes 27
High-Risk (Unsafe) Conditions and High-Risk (Unsafe) Behaviors 28
    High-Risk (Unsafe) Conditions 29
    High-Risk (Unsafe) Behavior 29
Exposure, Impact, or Energy Exchange 30
Illness, Injury, Property Damage, and Business Interruption 31

Contents   vii

        Cost of Accidental Loss 31
        Safety Management System (SMS) 32
        Summary 32

Chapter 4   Defining Safety Management Performance 33

        Safety Management 33
        Performance 33
        Safety Management Performance 33
        Safety Management Performance Measurements 33
           A Complex Challenge 34
        Types of Safety Management Performance Measurement
        Indicators 34
           Leading Measurements 35
           Lagging Measurements 35
           Soft Measures 35
           Hard Measurements 36
        Why Measure Safety Management Performance? 36
        Benefits of Safety Management Performance Measurement 36
        Safety Effort and Experience (SEE) 36
        Appropriate Mix of Measures 37
        Focus 37
        Management Review 37
        Against Standards and Objectives 37
           Internal Measurements 37
           External Measurements 38
        What Gets Measured, Gets Done 38
           Safety Is a State 38
           Too Simplistic 38
        If It Cannot Be Measured, It Will Not Be Done 39
        Risk Profile 39
           Risk Free 39
           Alarp 39
        Input Measurements (Positive Performance Indicators) 39
        Process Measurements 40
        Output Measurements 40
        Characteristics of Good Safety Management Performance
        Measurement Indicators 41
        The Safety Record 41
           Public Perception 41
           Differing Rules 41
           Safety Fear Factor 41
           World Safety Record 42
           Safety Verses Injury 42
           Paradigm 42
        Summary 42

PART II Safety Management Performance Measurements

Chapter 5 Pre-Contact, Contact, and Post-Contact Measurement 47

Pre-Contact, Contact, and Post-Contact Control 47
Pre-Contact Phase Control 47
    Positive Performance Indicators (PPIs) 47
    Pre-Contact Control 47
Pre-Contact Phase – Safety Performance Measurements
(Leading Indicators) 48
    Scoring Method 50
    Minimum Standard Detail 50
Measuring the Hazard Burden 51
Contact Phase of the Accident 51
    Contact Phase Control 51
Contact Phase – Safety Performance Measurements 52
    Trends 52
    Types of Energy Exchange 52
Accident and Near Miss Incident 52
    Case Study 53
Post-Contact Phase 53
Post-Contact Phase Measurement 53
    System Failure 54
    Accident-Based Metrics 54
Summary 54

PART III Measuring Lagging Indicators of Safety Management Performance

Chapter 6 Lagging Indicators of Safety Performance: Injuries, Illness, and Diseases 57

Introduction 57
Measures of Failure 57
Measures of Consequence 58
    Rear-View Mirror 58
    After the Event 58
    Consequences 58
    Outcomes 58
Advantages of Injury and Illness Data 58
    Disadvantages 59
Measurement of Control 59

Contents

- Guidelines to Recording Injury Experience 59
  - Employment 59
  - Occupational Injury 59
  - Occupational Fatality 60
  - Occupational Fatality Rates 60
  - Historic Measure 60
  - Golden Gate Bridge 60
  - Hoover Dam 61
  - High-Rise Construction Fatalities 61
  - Deadliest Project 61
  - World Cup 2022 Qatar 61
- Fatality-free Shifts 61
- Lost Time Injury 62
  - Temporary Total Disability (Lost Time Injury) 62
  - Shift Lost 62
  - Regular Job 62
  - The Acid Test 63
- Injury Frequency Rates 63
  - Universally Used 63
  - History 63
  - Comparison 63
  - National Figures 64
  - Legal Rating 64
- Lost Time Injury Frequency Rate (LTIFR) or Disabling Injury Frequency Rate (DIFR) 64
- Injury Incidence Rates 65
- Lost Time Injury Incidence Rates (LTIIR) or Disabling Injury Incidence Rate (DIIR) 65
- Total Injury Rate 66
- The Total Recordable Disease Frequency Rate 66
- OSHA Recordable Injury or Illness 66
- DART 66
- The Total Case Incident Rate 67
- Mine Safety and Health Administration (MSHA) Reportable Injury 67
- MSHA Incidence Rates 67
- RIDDOR (Reporting of Injuries, Diseases and Dangerous Occurrences) 68
- Injury Severity Rates 68
  - Workdays Lost Due to Work-related Accidents 68
- Average Days Charged per Disabling Injury 69
- Disabling Injury Index 69
- First Aid Defined 69
- Medical Treatment Defined 70
- First Aid Injury Rate 70
- Minor Injury Rate 70

Target Injury Frequency Rate (TIFR) 71
Analysis of High-Risk Behavior and High-Risk Workplace
Conditions 71
Exposure, Impact or Energy Exchange, Terminology 71
Energy Transfer (Exchange) Types 71
    Top three Causes 73
Agency 73
    Two Types of Agencies 73
Agency Part 75
    Agency Trends 75
Analysis of Body Part Injured 75
Other Safety Performance Metrics 75
Road Safety Performance Measurements 75
Fatal and Severe Injuries 76
Aircraft Accidents 76
Costs of Injuries and Illnesses 76
Conclusion 76
Summary 76

Chapter 7   Underreporting of Injuries, Illnesses, and Damage 78

Manipulation of Injury Definitions 78
Creative Bookkeeping 78
Underreporting of Injuries 79
    United States Government Accountability Office 79
    Hidden Tragedy 79
The Safety Fear Factor 80
Imbedded Culture 81
    Injured Workers Under Pressure 81
International Culture 81
No Blood – No Foul 81
    Near Miss Incidents 82
    Reporting Property Damage Accidents 82
    Safe Space 82
Safety Bribery 82
    Injury Free Bonuses 82
    The Root of Evil 83
Safety Incentive Schemes 84
Safety Awards 84
Injury-Free Periods 84
False Impression 84
Based on Actions 84
Safety Gimmicks 85
Safety Publicity Boards 85
    Incorrect Terminology 85
    Evil Record 85

Contents   xi

                Zero Harm 86
                    A Goal 86
                    An Objective 86
                    Strategy Must Be Defined 86
                    The Health and Safety Management System (SMS) 86
                Campaigns 87
                    Paper Cut Injury? 87
                Conclusion 87
                Summary 87

Chapter 8   Awards Based on Lagging Indicators 89
                The National Occupational Safety Association
                (NOSA) 89
                    Collection of Injury Statistics 89
                    NOSA Award Plan 89
                National Safety Council (US) Awards 90
                    Superior Safety Performance Award 90
                    Significant Improvement Award 90
                    Perfect Record Award 90
                    The Million Workhours Award 90
                Mining Awards 91
                The NOSA 5-Star Safety and Health Management
                System 91
                    Star Grading Awards 91
                    Safety Effort and Experience 91
                Occupational Safety and Health Administration (OSHA)
                Voluntary Protection Program (VPP) 92
                The VPP Star Program 92
                Summary 93

Chapter 9   Lagging Indicators of Safety Performance: Damage,
                Fires, and Interruption 94
                Accident Outcomes 94
                    Unintended Events 95
                Bridging the Gap 95
                Accidental Property Damage 95
                Accident Ratios 95
                    Accident Ratio Study 96
                    Other Ratios Researched 96
                    Criticism of Accident Ratios 97
                    Investigation and Recording 98
                    Risk Matrix 98
                    Calculating the Accident Ratio 99
                    Barriers to Reporting of Downgrading Events 99

  Amnesty 100
    Vital Statistics 100
  Calculating an Accident Ratio 100
  Costing 102
  Measuring Property Damage Accidents 102
  Fire Damage Accidents 103
  Environmental Harm 103
    Exxon Valdez Oil Spill 103
  Other Major Accidents 103
  Totally Hidden Costs 103
  Summary 104

Chapter 10 Cause of Injury or Damage: Transfer of Energy 105

  Impact, Exposure, or Energy Transfer Types 105
  Energy Release Theory 105
  Energy Exchange – Not Accident Types 106
    Energy Transfer Analysis 106
  Agency 107
    Agency Part 108
  Contact Control 108
  Summary 109

Chapter 11 Total Cost of Risk 111

  Accident Costs 111
    Increased Premiums 111
  The Experience Modification Rate (EMR) 111
    Incidental Costs 112
  Hidden Accident Costs 112
    Totally Hidden Costs 113
  Iceberg Effect 113
    Minimizing Losses 113
    Example 113
  Main Motivation 114
    Profit-Driven 114
    Cost Reduction 114
  Fines 115
    Cost of Non-Compliance 115
    Increased Fines and Penalties 116
  Highest Ever 116
  Major Accidents 116
  Cost–Benefit 117
  Life's Value 117
  Reputation 117
    Severe Repercussions 117

White Paper 118
Total Cost of Risk 118
Summary 118

PART IV Leading Safety Management Performance Indicators

Chapter 12 Safety Management Control 121

Leading Safety Management Performance Indicators
(Measurements of Control) 121
Definition 121
Safety Controlling 121
Risk-based, Management-led, Audit-driven Safety
Management System (SMS) 121
    Measurable Performance Standards 123
    Safety Authority 124
    Safety Responsibility 124
    Safety Accountability 125
    System Standards 126
    Measuring Performance 126
    Critical Task Observation 126
    External Audits 127
    Internal Audit 127
    Accident Investigation 128
    Commendation 129
    Recognition for the Achievement of Proactive
    Objectives 129
    Management 129
Conclusion 129
Summary 129

PART V Examples of Positive Performance Safety Indicators (PPSIs)

Chapter 13 Near Miss Incidents as a Measurement of Safety
Performance 133

Significance of Near Miss Incident Recording
and Measurement 133
    Not Just Quantity 133
Leading Performance Indicators 133

Near Miss Incidents (Close Calls) 134
    Iceberg Effect 134
The Accident Ratio 134
    Large Numbers 135
Benefits of Near Miss Incident Reporting 135
Measurement of Near Miss Incident Reporting 135
    Formal Reporting 135
    Near Miss Incident Report Record Log 136
    Hazard Reporting System 137
    5-Point Checklist 138
    Safety Reporting Hotline 138
    Informal Reporting 138
    Risk Ranking 139
    Measurement Criteria 139
    Calculations 140
    Monthly Totals 140
    Rectification System 140
    Feedback 141
    Setting Targets 141
Example Metrics 141
Proactive Safety Performance Measurement 141
Summary 142

Chapter 14 High-Risk Behavior (Unsafe Act) and High-Risk
Conditions (Unsafe Conditions) 144

Hazards 144
High-Risk Behavior (Unsafe Act) 144
High-Risk Workplace Condition (Unsafe Condition) 144
Immediate Energy Exchange Causes 144
    Event and Consequence 144
Accident Ratio 145
Loss Causation Model 145
Accident Investigation 145
    Metric 146
High-Risk Behaviors (Unsafe Acts) 146
    A Safety Myth 146
    High-Risk Behavior Is Not a Near Miss Incident 147
    A Complex Situation 147
Human Failure 148
    Inadvertent Failure 148
    Deliberate Failure 148
    Exceptional Failure 148
    Active and Latent Failures 148
    Errors 148
High-Risk Workplace Conditions 150

Contents

    Identifying and Measuring Hazardous Situations 151
    Safety Observation Program 151
        Appointed Observer(s) 151
        Checklist 151
        Time of Observation 151
        Observations Process 151
        Safety Observation Card 152
        Observations 152
        The Observation Process 152
        Discussions 153
        Record Findings 153
        Tracking System 153
        Observation Metrics 153
        Report 153
    Incident Recall 153
        Goal of Incident Recall 154
    Summary 154

Chapter 15 Health and Safety Inspections 156

    Key Performance Indicators 156
    Inspections 156
    Purpose of Safety Inspections 157
    Types of Hazards 157
        Safety Hazards 157
        Biological Hazards 157
        Chemical Hazards 158
        Ergonomic Hazards 158
        Physical Hazards 158
        Psychosocial Hazards 158
    Hazard Identification 158
        Hazard Classification 158
        Hazard Profiling 158
    Measuring Against Standards 159
        Checklist 159
    Inspection Guidelines 159
    Types of Inspections 160
        Health and Safety Representatives' Inspections 160
        Risk Assessment Inspection 160
        Legal Compliance 160
        Informal Walk About 161
        Planned Inspections 161
        Safety Observation 161
        Safety Department Inspection 161
        Housekeeping Competition Inspection 161
        Safety Survey 161

          Safety Audit Inspection 162
          Specific Equipment Inspections 162
          Other Inspections... 162
     Conclusion 163
     Summary 163

Chapter 16 Hazard Identification and Risk Assessment.................................... 164

     The Purpose of Hazard Identification and Risk Assessment 164
     Sources of Hazards and Hazard Burden 164
     Measuring the Hazard Burden 164
          Ongoing Assessments.. 165
     Hierarchy of Control 165
     Risk Assessment.. 165
     Risk Register 166
     Application of the HIRA Processes 166
     Audit of an HIRA Process 166
     Audit Protocol 168
     Key Safety Performance Indicator 168
     Summary 168

Chapter 17 Health and Safety Committee Meetings 170

     Measurable Management Performance 170
     A Health and Safety Committee 170
          Joint Health and Safety Committee.. 170
          Health and Safety Representatives' Committee 170
     Functions of Health and Safety Committees 171
     Leadership 171
     The Committee System 171
     Meeting Agenda 171
     Meeting Minutes 172
     A Vital SMS Component... 172
     Summary 173

Chapter 18 Health and Safety Representatives Appointed 174

     Health and Safety Representatives.. 174
     Safety Management Principles.. 174
          Principle of Safety Communication 174
          Principle of Safety Participation.. 174
          Principle of Safety Recognition... 174
     Who Should Be Appointed?.. 175
     Training 175
     Duties of Health and Safety Representatives 175
     The Health and Safety Representative System............................... 176

Contents   xvii

    Positive Performance Indicator 176
    Summary 176

Chapter 19 Safety Perceptions Surveys 177

    Safety Perception Surveys... 177
      Hazard Identification 177
      Improve Health and Safety 177
    What Is the Purpose of a Safety Perception Survey?...................... 177
    Leading Indicator 178
    Safety Perception Survey Objectives 178
    Areas to Cover in the Survey 178
    Survey Instrument and Questions 179
      The Likert Scale 179
      Benefits of Using Likert Scale Questionnaires......................... 179
    Feedback 180
    Reevaluation 180
    Summary 180

Chapter 20 Employees Trained in Health and Safety 181

    Safety Management Principles 181
    Accident Root Causes 181
    Induction Training 181
    Annual Refresher Training 182
    First Aid Training 182
      First Responder Training 182
    Health and Safety Coordinators 182
    Hazard Communication Training 182
    Critical Task Training 183
    Health and Safety Representative Training 183
    Competency-Based Training 183
    Technical Health and Safety Training 183
    Formalized 184
      Accident and Incident Recall 184
      Key Performance Indicator 184
    Summary 184

Chapter 21 Safety Toolbox Talks and Task Risk Assessments.......................... 186

    Purpose of Toolbox Talks 186
      Topics 186
      Occupational Safety and Health Administration 186
      Structured 187
      Meaningful 187
      Attendance Sheet 187

Task Risk Assessments 187
  Key Performance Indicator 187
Summary 188

Chapter 22 Quality of Accident Investigation Reports 190

Accident Investigation 190
Quality of Investigations 190
Scoring an Accident Investigation Report....................................... 190
Evaluation Checklist... 191
  General Information (10 Points) 191
  Risk Assessment (5 Points)... 191
  Description of the Event (15 Points)... 191
  Immediate (Cause of Energy Exchange) Cause Analysis
  (7 Points)... 192
  Root Cause Analysis (8 Points) 192
  Sketches/Pictures/Diagrams... 192
  Risk Control Measures (Remedies) (30 Points).......................... 192
  Investigation Commencement (20 Points).................................. 192
  Signatures (5 Points) 192
Has the Accident Investigation Been Effective? 193
Summary 194

PART VI The Safety Management System (SMS) Audit as a Safety Management Performance Measurement Tool

Chapter 23 The SMS Audit as a Safety Management Performance
Measurement Tool 197

Safety Management Performance Measurement 197
What Is an SMS Audit?... 197
  Not a Safety Inspection 197
  Proactive Approach 198
  Reactive Versus Proactive Measurement..................................... 198
  Subjective Versus Objective 198
Benefits of SMS Audits 198
  A Learning Opportunity 199
Types of SMS Audits... 199
  Baseline Audit 199
  Records and Verification 200
  Who Should Conduct the Baseline Audit? 200
  Internal SMS Audits 200
  Independent Audit (External Third Party) 200

Compliance Audit 200
Centralized and Decentralized SMS Coordination 201
Auditable Organization 201
   Solution 201
Audit Requirements 202
   Review of Element Standards 202
   Document Control System 202
   Incomplete Verification 202
   Audit Frequency 202
   Verification 203
SMS Audit Protocol 203
   The Audit Protocol 203
Who Should Conduct Audits? 203
   Auditor's Training 203
   Guidelines for Auditors 205
   Auditor's Experience 205
   Lead Auditor 205
   Questioning Techniques 205
   Compliment 206
Summary 206

Chapter 24 Example Leading Safety Key Performance
Indicators 207

Objective 207
Responsibility and Accountability 207
Requirements 207
Safety Committees 207
   Scoring Method (Maximum 20 Points) 208
   Alternate Score 208
Near Miss Incident Reports 208
   Scoring Method (Maximum 10 Points) 209
   Alternate Score 209
Safety Observations Reported 209
   Scoring Method (Maximum 10 Points) 209
   Alternate Score 210
Plant (Workplace) Inspections Completed 210
   Scoring Method (Maximum 10 Points) 210
   Alternate Score 210
Safety Management System Audit Results 210
   Scoring Method (Maximum 5 Points) 211
Fire, Evacuation, or Emergency Drills Held 211
   Scoring Method (Maximum 5 Points) 211
Number of Safety Toolbox Talks Held 211
   Scoring Method (Maximum 10 Points) 211
   Alternate Score 211

Employees Attending Health and Safety Training
Programs or Workshops 212
    Scoring Method (Maximum 5 Points) 212
Number of Safety Representatives Appointed
and Active 212
    Scoring Method (Maximum 15 Points) 212
    Alternate Scores 212
Task or Site Risk Assessments Conducted 212
    Scoring Method (Maximum 10 Points) 213
Score Sheet 213
Summary 213

References 215

Index 217

About the Author

Ron C. McKinnon, CSP (1999–2016), is an internationally experienced and acknowledged safety professional, consultant, author, motivator, and presenter. He has been extensively involved in safety research concerning the cause, effect, and control of accidental loss, near miss incident reporting, accident investigation, safety promotion, and the implementation of health and safety management systems for the last 50 years.

He received a National Diploma in Technical Teaching from the Pretoria College for Advanced Technical Education, a Diploma in Safety Management from the Technikon SA, South Africa, and a Management Development Diploma (MDP) from the University of South Africa in Pretoria. He received a master's degree in health and safety engineering from the Columbia Southern University.

From 1973 to 1994, Ron C. McKinnon worked at the National Occupational Safety Association of South Africa (NOSA), in various capacities, including General Manager of Operations and then General Manager Marketing. He is experienced in the implementation of health and safety management systems (SMS), SMS auditing, near miss incident and accident investigation, and safety culture change interventions.

From 1995 to 1999, Ron C. McKinnon was a safety consultant and safety advisor to Magma Copper and BHP Copper North America, respectively. In 2001, Ron spent two years in Zambia, introducing world's best safety practices to the copper mining industry. After leaving Zambia, he was recruited to assist in the implementation of a world's best class safety management system at ALBA in the Kingdom of Bahrain.

After spending two years in Hawaii at the Gemini Observatory, he returned to South Africa. Thereafter, he was contracted as the Principal Health and Safety Consultant to Saudi Electricity Company (SEC), Riyadh, Saudi Arabia, to implement a world's best practice safety management system, throughout its operations across the Kingdom involving 33,000 employees, 27,000 contractors, 9 consultants, and 70 safety engineers.

Ron C. McKinnon is the author of *The Cause, Effect and Control of Accidental Loss* (2000), *The Cause, Effect and Control of Accidental Loss, Second Edition* (2023), *Safety Management, Near Miss Identification, Recognition and Investigation* (2012), *Changing the Workplace Safety Culture* (2014), *Risk-based, Management-led, Audit-driven Safety Management Systems* (2016), *The Design, Implementation and Audit of Occupational Health and Safety Management Systems* (2020), and *A Practical Guide to Effective Accident Investigation* (2022), all published by CRC Press, Taylor & Francis Group, Boca Raton, USA. He is also the author of *Changing Safety's Paradigms*, published in 2007 by Government Institutes, USA, and the second edition published in 2018.

Ron C. McKinnon is a retired professional member of the American Society of Safety Professionals (ASSP) and an honorary member of the Institute of Safety Management (South Africa). He is currently a health and safety management system consultant, safety culture change agent, motivator, and trainer. He is often a keynote speaker at health and safety conferences and consults to international organizations.

Preface

Occupational health and safety in the workplace is a unique management science and provides many challenges to both the employees and management. The biggest challenge of all is keeping workers healthy and safe while they are at work. This means sending workers home at the end of the day in the same condition that they were in when they came to work. Considering the many hazards that workers could encounter in workplaces, this is indeed a challenge.

To gauge whether management is achieving this goal or not, some form of safety measurement or gauge is needed to indicate safety success or failure. These are referred to as safety metrics, which are measurements of safety management performance. In basic terms, they indicate whether an organization is safe or not so safe. Everything must be measured and so too must safety performance. What gets measured gets managed. Managing safety, however, is difficult if the wrong measures are used and management does not get a holistic view of the organization's safety performance.

Safety is traditionally defined as being injury-free, which is part of the problem. *Safety* in modern terms is the application of safety management methods, processes, and programs to reduce workplace risk to an acceptable level, which in turn will lead to fewer downgrading events and consequences such as workplace injuries.

Traditionally, safety performance was measured by the number and frequency of workplace fatalities and serious injuries. Unfortunately, archaic research blamed workers' unsafe acts for accidental injuries, and this led to the misconception that an injury was reason for punishment.

This punitive reaction to workplace fatalities and injuries seems to have been passed down through the ages and has created a modern-day safety fear factor whereby workers are reluctant to report injuries, and management does everything to pass blame for injuries onto the workers involved. This skews accurate safety performance measurement, as many injuries are hidden or otherwise disguised as minor occurrences so as not to affect the performance figures.

Because of the flaws that exist in using injuries as a sole measure of safety performance, new, meaningful indices are needed. These measurements should measure leading, proactive safety actions, activities, and efforts, and not just the results of accidents. These positive, leading measurements are based on work being done to combat accidental loss, and are realistic, manageable actions and activities. These are the inputs that occur within the health and safety management system (SMS) such as the number of risk assessments conducted, numbers of employees exposed to formal safety training, safety committee meetings held, and SMS audit scores.

By measuring leading safety efforts, management will be able to accurately measure safety performance. Proactive, realistic, achievable, and manageable safety goals can be set, and performance measured. Basing goals on accident results is inaccurate and unreliable, as once the accident sequence has been set in motion, we have no control of the outcomes.

While safety experience, in the form of injury and severity rates, is also important, a combination of both leading and lagging measurements will give management an overall picture of safety performance that uses both indicators.

Part I

Introduction to Safety Management

1 Safety Management

INTRODUCTION

Profits are an important measurement of the organization's success. Units produced, units sold, and return on investments are the universally accepted measurements of a company's performance. Other measurements are also involved in the gauging of the company's business success or failure.

CONSEQUENCE-BASED MEASUREMENTS

Measurement of safety performance is a guiding factor as to how well an organization manages health and safety in the workplace. It indicates if the organization is performing well or failing in its safety efforts. Unfortunately, due to the history of safety in the workplace, these measurements have often only been based on the consequences of loss producing events or the end result of accidents, which is typically the visible loss in the form of injury or disease to employees. Most safety performance measurements measure safety experience rather than safety effort.

NOT AN ACCURATE INDICATOR

Despite numerous recommendations by safety professionals and researchers, the use of loss measurements has prevailed and is universally understood, and widely used as the sole measure of safety performance. Many safety researchers have warned that the number and severity of injuries and occupational diseases experienced by an organization are not the accurate reflection of the safety efforts at a company. However, they are still widely used today.

In Chapter 13 of *The Measurement of Safety Performance,* W. Tarrants (1980) discusses major issues and identified safety problem areas. Concerning the use of injuries as a measurement, he states that:

> Measurements are needed that predict, not simply record, accident recurrences. Historical records of accidents mean nothing unless they can be used for prediction and control purposes. A measurement problem exists in that low injury frequency rates tell nothing about the potential for catastrophe. For example, one plant experienced a $45 million accident loss, with its walls covered with safety awards received based on a low frequency rate. Better measurement techniques are needed to identify loss potential.
>
> pp. 235–237

TRADITIONAL MEASUREMENT

After an accident, the resultant injury has traditionally been the main focus. Many safety professionals still focus on the injury to people as their main concern. The perception of safety performance is normally based on the number and degree of severity of injuries that occur in a workplace.

An injury is the only tangible proof that something has gone wrong with the process. It is also tangible proof of the consequences of either a high-risk work environment or behavior. Injuries are spectacular, instant, and graphic results of the consequences of an accident.

W. H. Heinrich (1959) argues:

> An injury is merely the result of an accident. The accident itself is controllable. The severity or cost of an injury that results when an accident occurs is difficult to control. It depends upon many uncertain and largely unregulated factors – such as the physical or mental condition of the injured person, the weight, size, shape, or material of the object causing the injury, the portion of the body injured, etc. Therefore, attention should be directed to the accident rather than to the injuries that they cause.
>
> p. 28

FALSE SENSE OF SECURITY

Organizations that experience fewer injuries than others are lured into a false sense of security by assuming that they are safer. Good safety controls are often *assumed* when an organization has low injury rates. Injuries are a poor indication of safety performance and are an even poorer indication of safety success. Injuries indicate a failure in the system and are more likely a measure of failure than of success.

SAFETY PERFORMANCE

As McKinnon (2000) states:

> The misconception that exists internationally is that a high number of injuries indicate "poor safety" and that an absence of injuries indicates "good safety." Quotations from authorities in this field will be given to show that injuries are used internationally to judge safety-management success. They are used for comparative purposes between organizations within the same state, between states, and even between organizations in different countries.
>
> p. 131

The tradition has always been to rank a company's safety performance as well as its management's safety performance by the number of injuries experienced over a specific period. Internationally, this precedent has been set. It would prove to be exceedingly difficult to convince managers that the injuries they are using to measure their safety management controls are, in actual fact, merely indications of either good or poor fortuity. This is proposed because once the event has been set in motion, we have no control over the outcome.

NO CONTROL

Once an accident sequence has been initiated, the outcomes are beyond our control. A brick falling off of an unguarded scaffold may result in serious injury to workers below. It may result in minor injury. It may miss all the workers, luckily, and fall harmlessly to the ground. If it falls to the ground without causing any injury, the event is mostly ignored. Normally, it is not considered in the safety performance equation. Should it cause serious injury however, then the injury becomes a statistic in the safety performance measurement. The question can be asked, "Are we measuring safety or luck?"

Dan Petersen (1998) talks on safety performance measuring systems as follows:

> In the early days of safety, accident measures, such as the number of accidents (injuries), frequency rates, severity rates, and dollar costs, were used to measure progress. Although it became clear long ago that these measures offer little help, they continue to be used today. Why should we consider other measures? "Results" measures nearly always measure only luck – unless they concern a huge corporation that generates thousands of injuries.
>
> p. 37

SHIFTING THE PARADIGM

The following chapters explain the parameters needed for realistic, meaningful, and accurate measures of safety performance. They show how pre-contact and proactive safety measurements tell management where they are going, rather than where they have been, with regard to safety performance.

CONCLUSION

An accurate measure of health and safety performance is vital to any organization. The measurement should be of such a nature that it gives management a holistic view of the health and safety efforts and results. For the health and safety profession, this may mean looking outside of the box and considering all metrics relating to health and safety at the workplace and not just the obvious indicators of the number of injuries or diseases experienced.

SUMMARY

Measurement of safety performance is a guiding factor as to how well an organization manages health and safety in the workplace. It indicates if the organization is performing well or failing in its safety efforts. Unfortunately, due to the history of safety in the workplace, these measurements have often only been based on the consequences of loss producing events or the end result of accidents.

An injury is the only tangible proof that something has gone wrong with the process. It is also tangible proof of the consequences of either a high-risk work environment or behavior.

Good safety controls are often *assumed* when an organization has low injury rates. Injuries are a poor indication of safety performance and are an even poorer indication of safety success. Once an accident sequence has been initiated, the outcomes are beyond our control. A brick falling off of an unguarded scaffold may result in serious injury to workers below. It may result in minor injury. It may miss all the workers, luckily, and fall harmlessly to the ground. If it falls to the ground without causing any injury, the event is mostly ignored. There is a need for realistic, meaningful, and accurate measures of safety performance. Pre-contact and proactive safety measurements tell management where they are going and injury rates tell management where they have been.

2 Workplace Health and Safety

WORKPLACE INJURY AND DEATH STATISTICS

According to the International Labor Organization (ILO) (2023):

> The ILO estimates that some 2.3 million women and men around the world succumb to work-related accidents or diseases every year; this corresponds to over 6000 deaths every single day. Worldwide, there are around 340 million occupational accidents and 160 million victims of work-related illnesses annually. The ILO updates these estimates at intervals, and the updates indicate an increase of accidents and ill health.
>
> **ILO website, Copyright © International Labor Organization (2023)**

According to the Health and Safety Executive (HSE) UK, key figures for Great Britain (2020/21):

> One million seven hundred thousand working people suffered from a work-related illness, 123 workers were killed at work during the same period, and 441,000 working people sustained an injury at work according to the Labor Force Survey. During the same period, 822,000 workers suffered work-related stress, depression, or anxiety and 470,000 workers suffered from a work-related musculoskeletal disorder. Updated estimates show the total cost associated with workplace injuries and ill health in Great Britain in 2010/11 to be some £13.8 billion (USD 15.6 billion).
>
> **HSE website (2023)**

Injury Facts 2022 (National Safety Council) states that:

The number of preventable work deaths increased 9% in 2021, totaling 4,472. In addition to preventable fatal work injuries, 718 homicides and suicides occurred in the workplace in 2021. These intentional injuries are not included in the preventable-injury estimates.

The increase in preventable work death in 2021 is partially a result of a 5% increase in the hours worked resulting from the economic recovery from the COVID-19 pandemic. The increase in hours worked does not fully account for the increase in work deaths. The preventable injury death rate of 3.1 per 100,000 workers is up from 3.0 in 2020. Work-related medically consulted injuries totaled 4.26 million in 2021.

During 2021 there were 4,472 preventable injury-related deaths, and the preventable injury-related death rate was 3.1 deaths per 100,000 fulltime workers. There were 4,260,000 medically consulted injuries during the same period.

The leading cause of work-related injuries and illnesses involving days away from work in 2020 is exposure to harmful substances or environments. Exposure to harmful substances or environments was previously the 6th ranked cause. The next two most prevalent causes of injury and illness involving days away from work are overexertion and bodily reaction, and slips, trips and falls. These top three causes account for more than 75% of all nonfatal injuries and illnesses involving days away from work.

NSC Website (2023)

SAFETY

There are many definitions of safety. Many people quote Webster's definition, which is: *The condition of being safe from undergoing or causing hurt, injury, or loss.* Other definitions include:

- *Safety is the identification and control of accidental loss before it occurs.*
- *Safety is freedom from danger, hurt, injury or loss.*

The definition of safety which will be used throughout this publication is: *safety is the control of accidental loss.* This means that safety is a control function, and it encompasses all forms of accidental loss, including injury, illness, equipment and property damage, business interruption, and environmental harm.

SAFETY IS A STATE

Safety is a state (the particular condition that someone or something is in at a specific time) in which hazards and conditions leading to physical, psychological, or physical harm are controlled in order to preserve the health and safety of individuals.

A SAFETY MEASURE

A safety measure is an action, procedure, intervention, or contrivance designed to lower the probability of the accidental occurrence of injury to people, loss of property, equipment, or the environment.

OCCUPATIONAL HYGIENE

Occupational hygiene is the science and art devoted to the anticipation, recognition, identification, evaluation, and control of environmental stresses arising out of a workplace which may cause illness, impaired well-being, discomfort, and inefficiency in employees or members of the surrounding community. Occupational hygiene is also described as the science dealing with the influence of the work environment on the health of employees.

OBJECTIVE

The objective of occupational hygiene is to recognize occupational health hazards, evaluate the severity of these hazards, and eliminate them by instituting control measures. Where the occupational health hazard cannot be eliminated entirely, occupational hygiene control methods aim to reduce exposure to the hazard and institute measures to reduce the hazard.

Occupational Hygiene Stresses

In a work environment, the employees could be exposed to numerous occupational hygiene hazards and stresses. They could include some of the following:

- Chemical hazards.
- Exposure to noise.
- Exposure to dusts.
- Exposure to steam.
- Exposure to smoke.
- Exposure to fuels.
- Exposure to heat extremes.
- Radiation exposure.
- Vibration.
- Ergonomic hazards, etc.

HAZARD

A *hazard* can be defined as: *a situation which has potential for injury, damage to property, harm to the environment or a combination of all three.* A hazard is a source of potential harm. High-risk behaviors and high-risk work conditions are examples of hazards.

The major hazard classifications are:

- Safety – slipping and tripping hazards, poor machine guarding, equipment malfunctions, or breakdowns.
- Biological – bacteria, viruses, insects, plants, birds, animals, humans, etc.
- Chemical – depends on the physical, chemical, and toxic properties of the chemical.
- Ergonomic – repetitive movements, improper workstation setup, etc.
- Physical – radiation, magnetic fields, pressure extremes (high pressure or vacuum), noise, etc.
- Psychosocial – stress, violence, etc.

POTENTIAL HAZARD

Hazards, which have the potential to cause an accident, are normally termed *potential hazards*. The hazards already exist; it is only the outcome which has the potential.

RISK

A *risk* can be defined as: *any probability or chance of loss*. It is the likelihood and severity of an undesired event occurring at a certain time under certain circumstances. The two major types of risks are *speculative* risks, where there is the possibility of both gain and loss, and *pure* risks, which offer only the prospect of loss.

HAZARD IDENTIFICATION AND RISK ASSESSMENT (HIRA)

The purpose of HIRA is to identify hazards and evaluate the probability of injury or illness arising from exposure to a hazard, with the goal of eliminating the risk, or using control measures to reduce the risk at the workplace.

The organization should perform ongoing documented HIRAs for the work at hand. The assessments should identify competencies required of the employees, as well as controls and barriers required to guard against identified hazards. The assessments should include identifying which company policies, standards, procedures, and processes apply to workplace situations. The organization's health and safety department may be consulted to establish a risk ranking for the various work processes.

A HIRA process include the following steps:

- Identify the hazards.
- Do a hazard classification exercise.
- Conduct a risk analysis considering probability, severity, and frequency.
- Allocate risk scores.
- Rank the risks according to the scores.
- Evaluate the risks.
- Compile a risk profile.
- Identify controls to be put in place, with consideration given to how effective and adequate the proposed controls would be.
- Review the assessment and update if necessary.

RISK ASSESSMENT

Risks cannot be properly managed until they have been assessed. The process of risk assessment can be defined as: *the evaluation and quantification of the likelihood of undesired events and severity of injury and damage that could be caused by the risks.* It also involves an estimation of the results of undesired events.

One of the biggest benefits of risk assessment is that, via risk evaluation, it will indicate where the greatest gains can be made with the least amount of effort and which activities should be given priority. The safety management system (SMS) now has a prioritization system based on sound risk assessment practices.

COMPONENTS

Risk assessment has four major components:

1. Hazard identification.
2. Risk analysis.
3. Risk evaluation.
4. Risk control.

A risk matrix (Figure 2.1) is one method to analyze the risk of hazards and enable a priority for risk mitigation.

Workplace Health and Safety

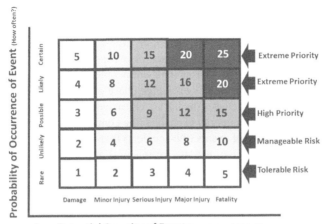

FIGURE 2.1 A risk matrix.

HIERARCHY OF CONTROL

There should be multiple layers of controls protecting employees from hazards and associated risks. These controls are measurable and would form part of the bouquet of safety management performance measurements discussed in other chapters. The types of controls could include:

- Elimination – which would mean changing the way the work is to be done.
- Engineering controls – such as the re-design of a worksite, equipment modification, and tool re-design.
- Administrative controls – such as a change of work methods and rescheduling of work.
- Health and safety system controls – such as health and safety policies and standards, work procedures, training, and personal protective equipment (PPE) (SMS elements).

An ongoing SMS will entail the ongoing implementation, application, and monitoring of one or more, or a combination of these remedies, to prevent accidents.

BUSINESS INTERRUPTION

A business interruption is a temporary delay in the work process because of an accident.

ACCIDENT

An accident is defined as: *an accident is an undesired event, which results in harm to people, damage to property or loss to process.*

Some sources incorrectly define an accident as: *that occurrence in a sequence of events that may produce unintended injury, illness, death, and/or property damage.*

Accident refers to the event, *not* to the result of the event. The *occurrence* is one segment within the accident sequence, which infers that one *occurrence* in the accident sequence causes the injury, death, or illness, and this is correct, but that *occurrence* is actually the exposure, impact, or exchange of energy segment of the accident sequence. That is what causes injury to people or damage to property. This is normally referred to as the *energy exchange* phase of the accident sequence.

It is not the accident that causes the harm. The accident is the sequence of events. It is the exposure, impact, or energy exchange (the contact and energy exchange segment) within the sequence of events (the accident) that causes the injury.

An accident is a sequence of events. An injury is a consequence of an accident. An accident is not an injury but may cause an injury.

OTHER DEFINITIONS

- *An accident is an undesired event often caused by unsafe acts and/or unsafe conditions and results in physical harm to persons and/or damage to property and/or business interruption.*
- *An accident is an unplanned uncontrolled event caused by unsafe acts and/or unsafe conditions and which results in harm to people or damage to property and equipment.*
- *An accident is the outcome of a series of activities, conditions, and situations and which ends in injury, damage, or interruption.*
- *An accident is an undesired event often caused by unsafe acts and/or unsafe conditions and results in physical harm to persons and/or damage to property and/or business interruption.*

From here on, this publication will describe an accident as: *an accident is an undesired event, which results in harm to people, damage to property or loss to process.* (Loss is experienced.)

NEAR MISS INCIDENT

Near miss incidents are also known as near misses, incidents, close shaves, warnings, or near hits. Other familiar terms are *close calls*, or in the case of moving objects, *near collisions*. Near miss incidents have also sometimes been termed *near hits* by some writers.

A near miss incident is defined as: a*n undesired event, which, under slightly different circumstances, could have resulted in harm to people, and/or property damage, and/or business disruption.*

Other definitions include:

- *An accident with no injury or loss.*
- *An event that narrowly missed causing injury or damage.*
- *An incident where, given a slight shift in time or distance, injury, ill-health, or damage easily could have occurred, but didn't this time around.*

The definition that will be used throughout this publication is: *an undesired event, which, under slightly different circumstances, could have resulted in harm to people, and/or property damage, and/or business disruption.*

Near miss incidents can also be defined *as: close calls that have the potential for injury or property loss.* Most accidents can be predicted by close calls. These are accidents that almost happened or possibly did happen but simply didn't result in an injury this time around. In fact, all the stages of the accident were present in the correct sequence except for the exchange of energy segments that would have caused injury, damage, loss, or a combination thereof.

ACCIDENT/INJURY/INCIDENT CONFUSION

There is still confusion as to what an accident is, what an injury is, and what a near miss incident is. Because of the archaic definition of the energy exchange, or impact, as *accident*, many still confuse the three terms. For clarity:

- An *injury* is personal harm resulting from an event called an accident (a consequence).
- An *accident* is a series of small blunders or a sequence of events that end up in a loss to people (injury), damage to property, business interruption, or environmental harm (an event).
- A *near miss incident* is also an undesired event which, under slightly different circumstances, could have resulted in harm to people, damage to property, business interruption, or environmental harm (an event with no loss but potential for loss).

The word *incident* is often used to describe an *accident,* but near misses were in the past also termed *incidents,* and this has created confusion with many now referring to accident/incident. When asking the question at international safety conferences, "Who is confused between the terms accident and incident?" I was met with a forest of hands. I estimated that about 90% of the attendees admitted that they were confused and did not fully understand the difference between an accident and a near miss incident.

CLARITY

An *accident* is a sequence of events that causes loss. A *near miss incident* is also a sequence of events that could have caused a loss under different circumstances. An *incident* describes a multitude of events. This publication will refer to accidents (loss-causing event) and near miss incidents (no loss but potential exists).

INJURY, DISEASE, OR ILLNESS

An injury is one of the many possible outcomes of an accident and is the most prominent consequence. It is tallied and used as a measure of safety effort, failure, or success.

An (accidental) injury is defined as: *the bodily hurt sustained as a result of an accidental contact. This includes any illness or disease arising out of normal employment.*

HIGH-RISK BEHAVIOR (UNSAFE ACT)

High-risk behavior is defined as: *a high-risk behavior is a departure from a normal accepted or correct work procedure, which reduces the degree of safety of that procedure.*

HIGH-RISK CONDITION (UNSAFE CONDITION)

High-risk workplace conditions are defined as: *a high-risk workplace condition is any physical condition which constitutes a hazard, and which may lead to an accident if not rectified.*

IMMEDIATE ACCIDENT CAUSES?

High-risk behaviors and high-risk workplace conditions are incorrectly termed the immediate *causes* of accidents. Correctly termed, they are the immediate cause of the *exposure, impact, or exchange of energy.*

ACCIDENT SEQUENCE OR LOSS CAUSATION SEQUENCE (DOMINO SEQUENCE)

An accident sequence is a model depicting the events leading up to an accidental exposure, impact, contact, exchange of energy, and resultant loss. The causation model is a chain reaction or series of events, which is often compared to the fall of a row of dominoes, giving the *domino effect* or *knock-on effect*, which is common to most accidents.

ACCIDENT CAUSATION THEORIES

There are many accident causation theories. Some are depicted by simpler linear models, and others by more complex models and advanced theories. A common thread is that accidents (the event) are caused and can be prevented. This excludes exceptional events such as the forces of nature and natural disasters.

LOSS AND NO LOSS

Few of the accident causation theories, however, can explain why some undesired events result in a loss, and similar events end up with no loss. The events in some cases are identical yet some cause loss and others don't. Many are now beginning to question if the outcomes of an undesired event are fortuitous. That means the difference between an accident that causes a loss and a near miss incident is fortuitous.

A BASIC LOSS CAUSATION SEQUENCE

The failure to identify hazards, assess and control risks triggers off poor controls in the form of an inadequate SMS. This causes personal and job factors known as the accident root causes. These accident root causes create a climate that breeds high-risk behaviors and high-risk conditions. The stage has been set and invariably the next event is either an exposure, impact, or contact with a source of energy, leading to a loss, or a near miss incident with no loss. The outcome is hard to predict and is sometimes the result of good or bad fortune.

According to McKinnon (2000):

> It is therefore concluded that there is no logical explanation why some high-risk behavior ends up as accidents (with loss) and the same high-risk behavior, under slightly different circumstances, ends up as a near miss incident (with no loss). The difference is determined by fortuity. This creates a dilemma when measuring safety management performance by using loss statistics only.
>
> p. 79

According to Safeopedia:

> For the purposes of safety statistics, accidents which cause harm and near misses are sometimes grouped together as a single category. This is because the outcome of an accident (whether it results in harm) is often a matter of luck.
>
> NSC Website (2023)

Mark Rothstein (1998) puts it this way, "On the other hand, the absence of an accident doesn't mean there was no violation – it may only reflect the employer's good fortune" (p. 135).

In the publication, *A Guide to Measuring Health and Safety Performance*, the Health and Safety Executive (HSE) (UK) (2022) states:

> Whether a particular event results in an injury is often a matter of chance, so it will not necessarily reflect whether or not a hazard is under control. An organization can have a low injury rate because of luck or fewer people exposed, rather than good health and safety management.
>
> p. 5

SHIFT IN TIME OR POSITION

Some define the Luck Factor as being in the wrong place at the right time. Some explain it as a result of timing, positioning, circumstance, unexplained differences, and other factors that were not planned nor could be explained.

Experience has shown that there is always an element of luck as to whether inadequately controlled hazards actually result in harm. Good performance in this quadrant may indicate that an organization has been lucky. However, one cannot rely on luck. Future performance may not be so good, and more attention to health and safety inputs is required to maintain the performance.

OSHA (2015) defines a near miss as:

An incident in which no property was damaged, and no personal injury was sustained, but where, given a slight shift in time or position, damage or injury easily could have occurred.

NARROWLY AVOIDED

Others explain a near miss incident as an unintentional incident that could have caused damage, injury, or death but was narrowly avoided. Yet they do not explain why the event was narrowly avoided.

Another way to put it is that, on the other hand, the absence of an accident does not mean there was no violation – it may only reflect the employer's good fortune.

H.W. HEINRICH

In 1931, H.W. Heinrich (Heinrich et al., 1959) presented a set of theorems known as the axioms of industrial safety. His fourth axiom reads:

The severity of an injury is largely fortuitous, the occurrence of the accident that results in injury is largely preventable.

LOSSES

The consequences of an accident could be losses such as injury, disease or illness, property damage, equipment damage, business interruption, product and production loss, environmental harm, and other indirect losses and disruptions.

A loss is an unplanned waste of any resource, be it injury, time, damaged product, or loss of process. A loss is a preventable waste of a resource.

TYPES OF LOSSES

Accidents result in some form of loss. This could be in the form of fatalities, injuries, occupational diseases, product and equipment damage, or business interruption. A near miss incident results in no loss. It should, however, be ranked for potential loss and probability of recurrence. Under different circumstances, it may have resulted in a loss.

The three main categories of loss are:

- Direct losses (Insured).
- Indirect or hidden losses (Uninsured).
- Totally hidden losses.

These are also referred to as insured and uninsured losses.

DIRECT LOSSES (INSURED LOSSES)

Direct losses, such as medical treatment are normally covered by insurance. In most instances, worker's compensation insurance pays for certain medical costs arising out of the accident.

MEDICAL

Modern workers' compensation laws provide fairly comprehensive and specific benefits to workers who suffer workplace injuries or illnessess. Benefits vary, but often include medical expenses, death benefits, lost wages, and vocational rehabilitation.

The National Safety Council (NSC) (US) publication *Injury Facts 2022* states that:

> The total cost of work injuries in 2021 was $167.0 billion. This figure includes wage and productivity losses of $47.4 billion, medical expenses of $36.6 billion, and administrative expenses of $57.5 billion. This total also includes employers' uninsured costs of $13.8 billion, including the value of time lost by workers other than those with disabling injuries who are directly or indirectly involved in injuries, and the cost of time required to investigate injuries, write up injury reports, and so forth. The total also includes damage to motor vehicles in work-related injuries of $5.4 billion and fire losses of $6.3 billion.
>
> NSC Website (2023)

COMPENSATION

Compensation for lost wages, permanent disability, or death as a result of a work-related accident is also normally covered by insurance. This varies from country to country. The compensation paid is a minimal amount and does not compensate for totally hidden costs such as pain and suffering and long-term lifestyle changes.

PERMANENT DISABILITY

Ongoing compensation may be paid out to an employee who suffered a permanent disability as a result of an accident. In some cases, this is a lump sum payment, and in other cases, it is an ongoing benefit paid to compensate for earnings the injured person would have earned before the accident. As a result of the injury, the injured worker may have to do a more menial, lower paying job than he or she had at the time of the accident.

REHABILITATION

In most cases, rehabilitation of the injured worker is also covered by workers' compensation. This would enable the injured worker to return to work in a position where they could recover from the injury or disease. This position would not expose the recuperating worker to stressful work situations that may aggravate their injury or disease. These positions are termed light duty and form a part of the company's return to work program, which is an element of their SMS. Rehabilitation would include medical treatment for the physical rehabilitation of injuries or the fitting of prostheses.

HEALTH AND SAFETY MANAGEMENT SYSTEM (SMS)

A health and SMS (safety system) is a formalized approach to health and safety management through the use of a framework that aids in the identification, control, and mitigation of safety and health risks. Through routine monitoring, an organization checks compliance against its own documented SMS, as well as legislative and regulatory compliance. It is a series of ongoing management processes. A well-designed and operated safety system reduces accidental loss potential and improves the overall management processes of an organization. Introducing a formalized SMS is the only way to change an organization's safety culture.

A Continuous Process

A SMS is a continuous, ongoing process that enables an organization to control its occupational health and safety risks and to improve its safety and health endeavors by means of continuous improvement of safety and health processes. The achievement of targets and goals must be sustained year in and year out. An organization will never be able to state that it has completed the safety management process. As with all processes, existing targets and goals need to be continually achieved and new goals and objectives will arise from time to time.

RISK-BASED, MANAGEMENT-LED, AND AUDIT-DRIVEN

Risk-based

The safety system must be a risk-based system. That means it must be aligned with the risks arising in the workplace. The emphasis on certain system elements will be different according to the hazards associated with the work and the processes used. There is unfortunately no SMS that will be ideal for all mines, industries, and other workplaces; therefore, available systems should be seen as a framework on which to build a risk-specific system for the specific industry type. The main aim of the system is to reduce risks; therefore, the system must be aligned with those risks.

Management-led

The key factor in safety and health management systems is management leadership. The safety system must be initiated and supported by senior management as well as line management. Only management has the authority and ability to create a safe and healthy workplace. This should be one of their prime concerns.

Safety systems that originate with, and which are maintained by the safety department, will have little effect on the organization. It is estimated that about 15% of a company's problems can be controlled by employees, but 85% can be controlled by management. Most safety problems are, therefore, management problems. If managers can manage the intricate and difficult concept of safety, then they will be able to manage other aspects of management easier, as managing safety enables them to be more effective.

Audit-driven

What gets measured usually gets done. Safety is an intangible concept and is traditionally measured after the fact – once a loss has occurred. The safety system must be an audit-driven system which calls for ongoing measurement against the standards and quantification of the results before an accident.

A safety system converts safety intended actions into proactive activities and assigns responsibility and accountability for those actions. Very similar to what a manager does with his subordinates. Each activity, included in the safety system elements can be scored on a 0–5 scale to determine whether it has been achieved or not. At the end of the day the entire system can be quantified by the score allocated. The safety system's effectiveness has been measured. The safety system elements and processes that scored less than full points are highlighted as areas that need improvement. Managers tend to pay more attention to processes that can be measured and quantified, and what gets measured gets done.

MANAGEMENT SAFETY FUNCTIONS

It is generally accepted that a manager's main safety functions are:

- Safety planning.
- Organizing for safety.
- Leading or directing the safety movement.
- Safety controlling.

All these functions entail the management of employees, materials, machinery, the environment, and processes. The four basic functions of safety management form a solid foundation for the SMS, and, if integrated into a manager's day-to-day activities, could provide for better management, leadership, and involvement in the SMS and its elements.

Safety Controlling

Safety controlling is the management function of identifying what must be done for health and safety, setting health and safety standards, inspecting to verify completion of work, evaluating conformance, and following up with safety action. This is the most important safety management function and is vital to the design, implementation, and maintenance of a successful SMS.

Risk-based, Management-led, Audit-driven SMS

Based on risk assessments, a manager lists and schedules the work needed to be done to create a healthy and safe work environment and to eliminate high-risk behavior of employees and high-risk workplace conditions. This would call for the introduction of a suitable risk-based, management-led, and audit-driven SMS based on the world's best practices and aligned to the risks of the organization. All SMS should be based on the nature of the business and be risk-based, management-led, and audit-driven. The management control function has 8 steps:

- I – Identify hazards and assess the risk.
- I – Identify the work to be done to mitigate and control the risks.
- S – Set standards of measurement.
- S – Set standards of accountability.
- M – Measure conformance to standards by inspection and review.
- E – Evaluate conformances and achievements.
- C – Correct deviations from standards.
- C – Commend compliance.

Step 1 – Identify the Hazards and Assess the Risk

The HIRA process will ensure that an undertaking has identified all the hazards, analyzed the risks, evaluated them, and ascertained which risk control methods to apply. These controls, which are measurable, would form the basis of the SMS.

Step 2 – Identify the Work to Be Done to Control and Mitigate Risks

Once the risks have been assessed, evaluated, and prioritized, it is now management's function to identify what work must be done, by whom, and by when to ensure the treatment of the risks. The risk assessment would have identified both physical and behavioral risks. Management can now implement certain control elements under the umbrella of the SMS to reduce the risks to a level as low as is reasonably practical.

The work to be done to reduce risks, in many instances, is very similar to the basic health and safety control activities required by industrial and mining legislation. Examples of the work that may need to be done are the following:

- HIRAs.
- Regular inspections of lifting gear.
- Legal injury and disease reporting.
- Safety induction training for new employees.
- Guarding of all machinery and pinch points.
- Providing and maintaining PPE.
- Hazardous work procedures and controls.
- Hazardous substance control.
- Controlling permit required work.
- Formal accident investigation procedures.
- Establishing policies and standards, etc.

Step 3 – Set Standards of Measurement

It is known that managers get what they want, and should management set health and safety standards, these standards are usually achieved by the organization. Setting standards of measurement clearly indicates how things must be in the work environment. Health and safety standards are measurable management performances. Standards must be in writing and should contain the following headings:

- Purpose.
- Resources.
- Responsibility.
- Legal requirements.
- General.
- Monitoring.
- Document control.

By setting standards of measurement, management defines the direction in which the organization moves. Should management set a standard for good housekeeping practices, this standard, which is measurable management performance, is then the way business is done in the future. What gets measured gets done. Standards give the company goals and directions and a definite focus on the end result. These standards can be of measurement and of performance and ask the questions, "What must the end result be?" and "What must be done by whom and by when?"

Standards of measurement should be set for all the elements, processes, and programs within the SMS. These could number up to eighty in a fully fledged SMS. These would include:

- Housekeeping.
- Stacking and storage.
- Hygiene monitoring.
- Environmental conformance.
- Health and safety committees.
- Safe work procedures.
- Risk assessments.
- Plant inspections, etc.

Step 4 – Set Standards of Accountability

The next step in the control process is the setting of standards of accountability. A standard of accountability indicates *who* will do *what* and *when*. Setting standards of accountability asks, "Who must do it and by when?" An example of setting standards of accountability is to make certain supervisors accountable for area inspections. Part of the appointment would be committing that person to a date for the completion of each inspection. A measurable management performance indicator has been created.

Authority, Responsibility, and Accountability

There is often confusion about authority, responsibility, and accountability, and it is opportune to define these concepts here:

- Safety authority is the total influence, rights, and ability of the post to command and demand safety.
- Safety responsibility is the safety function and duty allocated to a post or position.
- Safety accountability is when a manager is under an obligation to ensure that safety responsibility and authority are used to achieve both health and safety and legal safety standards.

Setting safety standards for accountability involves deciding *who* will do *what* and *when*. An example of a standard of accountability is that the site foreman has been made responsible for conducting daily site risk assessments.

1. Who? – The site foreman.
2. Will do what? – Will carry out task risk assessments.
3. When? – These risk assessments will be done on a daily basis.

One of the many safety myths is that "everybody" is responsible for safety. In fact, individuals can only be *responsible* for items and people over whom they have *authority* and can thus be held *accountable* for only those conditions and people over whom they have authority.

Step 5 – Measurement Against the Standard

This control function is when management measures what is actually happening in the workplace against the preset standards. To measure successfully, a walkabout inspection must be carried out, and employees doing these inspections should be aware of and familiar with the standards. One of the greatest failings in most SMS is insufficient or inadequate inspections.

Systems to enable ongoing measurement against standards are part of a SMS, and these could include the monthly inspections of local work areas by appointed health and safety representatives. Audit inspections are ideal measurement tools. Critical task observations also allow for measurement against standards. The setting of standards and constant measuring against those standards immediately identify the strengths and weaknesses of the SMS. Safety personnel should also conduct formal inspections on a regular basis and compare actual SMS processes and procedures with the standards. A checklist should always be used when doing these inspections, as it will serve as a constant reminder of what must be measured.

It should be emphasized here that this form of safety management control and the measurement phase of the control process are not merely measuring and comparing injury statistics with other companies or industries. This is a pure management function of measuring whether the organization is living up to the norms agreed to by management and employees in the form of SMS health and safety standards, and the health and safety policy statement.

Step 6 – Evaluate Conformance to SMS Standards and Achievements

The evaluation process is the quantification of the degree of conformance to the standards set. Evaluation of the achievement of standards is normally facilitated through the audit process. An SMS should be driven by audits. These regular audits systematically quantify the degree of compliance with standards. They evaluate the management work being done to combat losses. What gets measured gets attention, and consequently, the evaluation of compliance with safety standards gives an indication of what is being done and what is not being done. The quantification of safety control actions is far more reliable and significant than the measurement of safety consequences, which are largely fortuitous.

Step 7 – Correct Deviations from Standards

Corrective action is the safety management work that must be done to correct those activities that were not completely controlled. If any critical SMS element is evaluated at less than 100%, some action needs to be taken to ensure total conformance with standards. Management must do what it says it is going to do. The safety standards indicate what must be done. Deviations indicate that the safety objective has not yet been achieved.

Step 8 – Commendation for Compliance

One of the main failings in numerous safety processes and programs is the lack of commendation and recognition. Commendation should be given for the achievement of objectives. If a department meets and maintains the housekeeping standard, for example, the entire group should be commended. Commendation for pre-contact safety activities is far more effective than commendation for injury-free periods.

SUMMARY

There are many definitions of safety. Many people quote Webster's definition, which is: *the condition of being safe from undergoing or causing hurt, injury, or loss.* Other definitions include:

- *Safety is the identification and control of accidental loss before it occurs.*
- *Safety is freedom from danger, hurt, injury, or loss.*

The definition of safety which will be used throughout this publication is: *Safety is the control of accidental loss.* This means that safety is a control function, and it encompasses all forms of accidental loss, including equipment and property damage, business interruption, and environmental harm.

A *risk* can be defined as: *any probability or chance of loss.* It is the likelihood and severity of an undesired event occurring at a certain time under certain circumstances. The purpose of HIRA is to identify hazards and evaluate the probability of injury or illness arising from exposure to a hazard, with the goal of eliminating the risk or using control measures to reduce the risk at the workplace.

Near miss incidents are also known as, near misses, incidents, close shaves, and warnings or near hits. Other familiar terms are *close calls*, or in the case of moving objects, *near collisions*.

An *accident* is a sequence of events that causes loss. A *near miss incident* is also a sequence of events that could have caused a loss under different circumstances. An *incident* describes a multitude of events.

An accident sequence is a model depicting the events leading up to an accidental exposure, impact, contact, exchange of energy, and resultant loss. The causation model is a chain reaction or series of events, which is often compared to the fall of a row of dominoes, giving the *domino effect* or *knock-on effect*, which is common to most accidents.

3 Understanding and Analyzing Accidental Loss Causation

A SEQUENCE OF EVENTS

An accident is a sequence of events that results in accidental losses such as injuries or damage. This sequence of events is sometimes referred to as the domino effect, as once the initiating event is triggered, all the dominoes fall. The loss causation domino sequence, originally termed the accident sequence, was originally proposed by H.W. Heinrich in 1929 and has been revisited and updated by several safety pioneers. Many other accident causation theories exist, but some are complex and sometimes difficult to understand.

OTHER ACCIDENT CAUSATION THEORIES

Other loss causation models and theories based on the domino effect have also been proposed by many safety professionals, such as Frank E. Bird, Dan Petersen, and others. For the purposes of measuring safety management performance, a simplified loss causation model, which is a modified version of the Frank E. Bird theory, is proposed.

LOSS CAUSATION SEQUENCE

The loss causation sequence proposed here is that all forms of accidental loss are triggered by a failure to identify the hazards, analyze and evaluate the risks, and institute control measures in the form of a structured and sustained health and safety management system (SMS). This in turn leads to weaknesses in the management control system, which give rise to job and personal factors, commonly referred to as the root causes of accidents. These root causes prompt high-risk behavior to be committed and in turn allow high-risk conditions to be created.

Once this situation exists, there could be an accidental transfer of energy or not. A flow of energy but no energy transfer results in a near miss incident, which is commonly referred to as "nothing happened."

Should there be exposure, impact, contact, and energy exchange, the outcome could be injury, property damage, or business interruption or a combination of two or all three. If the exchange of energy causes personal injury, the severity of the resultant injury is fortuitous as it cannot be accurately predicted.

LOSSES

Losses normally occur because of accidents. An accident is an event that was neither planned nor can it be controlled once in motion. Once the first domino is triggered, the knock-on effect occurs, the end result of which is the loss in the form of injury, property damage, business interruption or a combination of these consequences.

Accidents are caused by a breakdown in the management control system (the health and SMS), and the result of every accident is some form of loss. The four main areas of loss are people, equipment, property, and the environment.

IMPORTANCE

The loss causation analysis is of vital importance to the safety management profession. It calls for a different way of looking at, measuring, and promoting the prevention of occupational injuries, damage, and diseases. The theory clearly demonstrates that traditional forms of safety measurement (lagging indicators) and the almost total disregard of positive, proactive leading indicators have to change before losses, such as the injury toll, can be reduced.

NEAR MISS INCIDENT

There is confusion as to the differentiation between *accidents*, *incidents*, and *near misses*. This occurred when the term *incident*, which was used to describe a *near miss event*, was changed to describe *accidents* and other happenings. Near misses lost their identity and were sadly neglected.

A *near miss incident* is defined as: *an undesired event, which under slightly different circumstances could have resulted in a loss.*

An *accident* is defined as: *an undesired event, which results in harm to people, damage to property or business interruption.* This means that accidents do result in losses, but near miss incidents do not result in any loss. They do, however, offer a warning as to the *potential* of loss occurring.

TRADITIONAL VIEWPOINT

In general, organizations do not normally acknowledge experiencing an accident until there is severe injury or illness to a person or persons. Most undesirable events do not end up in any loss at all. The majority of accidents cause property damage and minor injuries, and less than 2% of accidental occurrences result in serious injuries. Yet, it is traditionally serious injuries that are used as the main measurement of safety management performance.

According to the Cause, Effect and Control of Accidental Loss theory (McKinnon 2000), the end result (the energy exchange and nature of loss) of an undesired event is often swayed by fortuity or luck factors, over which an organization has little or no control. Once the sequence has been triggered, it is difficult to predict the outcomes.

Understanding and Analyzing Accidental Loss Causation

If the outcomes of an accident cannot be predicted, how can we then measure them accurately? The question is then, are we measuring control or luck?

FAILURE TO IDENTIFY HAZARDS AND ASSESS THE RISK

Accidents are caused by a sequence of events. A combination of circumstances and activities culminates in a loss. The loss may be an injury, damage, or business interruptions, or a combination thereof. The accident sequence is triggered by a failure to adequately identify the hazards and assess the risks they pose, which in turn causes a lack of or inadequate control in the form of a weak or non-existent SMS.

The hazard identification and risk assessment (HIRA) sequence is:

- Hazard identification.
- Risk analysis.
- Risk evaluation.
- Risk control.

If the risks posed by the enterprise are not identified and assessed, then controls in the form of structured health and SMS (SMS) are more than likely not in place to prevent the accident sequence from occurring (Figure 3.1).

WEAK OR NON-EXISTENT SAFETY MANAGEMENT SYSTEM

The second phase of the accident sequence is a lack of or inadequate control of risks. This lack of control could be because of a weak or non-existent SMS, no safety system, no safety system standards and controls, or non-compliance with the standards (Figure 3.2). This weakness leads to the root causes of accidents.

ACCIDENT ROOT CAUSES

The root causes of accidents are categorized as personal (human) and job (workplace) factors. They are the underlying reasons why high-risk behaviors are committed and why high-risk conditions exist. A personal factor could be a lack of skill, physical,

FIGURE 3.1 Failure to identify the hazards and assess the risks results in a weak health and safety management system. (From McKinnon, Ron C. 2000. *The Cause, Effect and Control of Accidental Loss (With Accident Investigation Kit)*. Boca Raton: Taylor and Francis. With permission.)

FIGURE 3.2 A weak or inadequate health and safety management system exists because of a failure to identify hazards and assess the risks fully. (From McKinnon, Ron C. 2000. *The Cause, Effect and Control of Accidental Loss (With Accident Investigation Kit)*. Boca Raton: Taylor and Francis. With permission.)

or mental incapability to carry out the work, a poor attitude, or a lack of motivation. Job factors could include inadequate procedures, poor maintenance, incorrect tools, or inadequate equipment (Figure 3.3). The root causes are what trigger the immediate causes of the exposure, impact, or exchange of energy, which are high-risk work conditions and high-risk behaviors.

HIGH-RISK (UNSAFE) CONDITIONS AND HIGH-RISK (UNSAFE) BEHAVIORS

The fourth domino in the accident sequence depicts high-risk behavior and high-risk conditions, commonly and incorrectly referred to as the immediate causes of accidents. They are the immediate cause of the energy exchange phase of the accident, not the accident itself.

A high-risk, or unsafe, act is defined as: *the behavior or activity of a person which deviates from normal accepted safe procedure*. A high-risk condition is defined as: *a hazard or the unsafe mechanical or physical environment*.

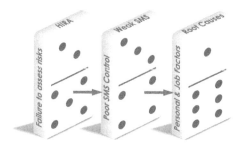

FIGURE 3.3 Accident root causes are created by a weak, failing, or non-existent health and safety management system. (From McKinnon, Ron C. 2000. *The Cause, Effect and Control of Accidental Loss (With Accident Investigation Kit)*. Boca Raton: Taylor and Francis. With permission.)

Understanding and Analyzing Accidental Loss

HIGH-RISK (UNSAFE) CONDITIONS

High-risk conditions are physical work conditions that are below accepted standards, contain hazards, and pose a risk of loss. This results in a high-risk area or an unsafe work environment. High-risk work conditions such as unguarded machines, cluttered walkways, poor housekeeping, inadequate lighting, and poor ventilation are responsible for a large percentage of all accidents and are a result of a weak or nonexistent safety system.

HIGH-RISK (UNSAFE) BEHAVIOR

High-risk acts are behaviors that put people at risk. This means that people are behaving contrary to accepted safe practices and are thus creating a hazardous, high-risk situation, which could result in accidental loss. In some instances, high-risk behaviors are committed knowingly, and sometimes they are because of poor communication or a lack of specific knowledge.

Examples

High-risk behaviors include working without authority, failure to warn somebody, rendering safety devices inoperative, or clowning and fooling around in the workplace. Numerous accidents are caused by high-risk behaviors. All high-risk behaviors and conditions can be traced to their root causes created by inadequate safety system control, because of a failure to assess and manage risk. All accidents have multiple causes, and an organization should not support certain philosophies that blame accidents entirely on human behavior (Figure 3.4).

Fortuity

Once high-risk behaviors or high-risk conditions are present, the stage has been set, and invariably, the next event is either an exposure, an impact, or an exchange of energy and a loss, or a near miss incident, depending on fortuity. The difference could be because of a slight shift in time or position.

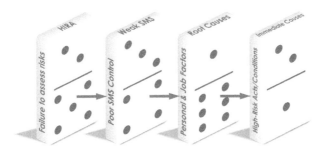

FIGURE 3.4 Root causes lead to high-risk behaviors and high-risk conditions. (From McKinnon, Ron C. 2000. *The Cause, Effect and Control of Accidental Loss (With Accident Investigation Kit)*. Boca Raton: Taylor and Francis. With permission.)

OSHA defines a near miss as: *an incident in which no property was damaged, and no personal injury was sustained, but where, given a slight shift in time or position, damage or injury easily could have occurred.*

Warnings

We have been aware of near miss incidents, close calls, narrow escapes, and warnings for a number of years. Safety pioneer Heinrich (1959) introduced his 10 axioms of industrial safety back in the 1930s. His third axiom reads as follows:

> The person who suffers a disabling injury caused by an unsafe act, in the average case, had over 300 narrow escapes from serious injury as a result of committing the very same unsafe act. Likewise, people are exposed to mechanical hazards hundreds of times before they suffer injury.

p. 21

Under Slightly Different Circumstances

Many agree that there is no logical explanation why some high-risk behaviors end up as accidents (loss) and the same high-risk behavior, under slightly different circumstances, ends up as a near miss incident (no loss). The difference in outcomes must be fortuitous, as one cannot control or accurately predict the consequence of a high-risk action or hazardous condition. The difference, therefore, between a near miss incident (close call) and an accident that produces a loss can only be ascribed to luck, good or bad fortune. This means that measuring safety management performance by using injury-based indicators is then measuring the end result of fortuity. The recording and measuring of high-potential near miss incidents are as important and more relevant than injury statistics.

EXPOSURE, IMPACT, OR ENERGY EXCHANGE

The exposure, impact, contact, or exchange of energy is the part of the accident sequence that is most closely associated with the loss. The exposure, impact, or energy transfer is what injures, damages, pollutes, or interrupts the business process (Figure 3.5).

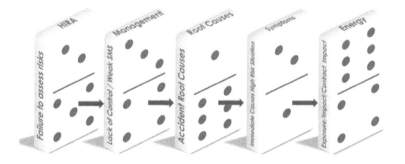

FIGURE 3.5 High-risk behaviors and/or high-risk conditions lead to an exposure, impact, or exchange of energy. (From McKinnon, Ron C. 2000. *The Cause, Effect and Control of Accidental Loss (With Accident Investigation Kit)*. Boca Raton: Taylor and Francis. With permission.)

Understanding and Analyzing Accidental Loss Causation

The exposure, impact, or exchanges of energy are also (incorrectly) referred to as *accident types*. A more apt description would be *energy transfers*. Examples of the classification of energy transfers are:

- Struck against (running or bumping into).
- Struck by (hit by a moving object).
- Fall to a lower level (either the body falls or the object falls and hits the body).
- Fall on the same level (slips and fall, tip over).
- Caught in (pinch and nip points).
- Caught on (snagged, hung).
- Caught between (crushed or amputated).
- Contact with (electricity, heat, cold, radiation, caustics, toxics, and noise).
- Overstress/overexertion/overload.

ILLNESS, INJURY, PROPERTY DAMAGE, AND BUSINESS INTERRUPTION

The transfer of energy caused by a high-risk behavior, high-risk condition, or combination thereof could result in illness, injury, damage to property and equipment, business interruption, or a combination of consequences (Figure 3.6).

COST OF ACCIDENTAL LOSS

All accidents cost money. Whether the outcome of an accident is a fatality, injury, property damage, or business interruption, financial losses are incurred. Most costs after an accidental loss are hidden. Many are difficult to calculate but are losses, nevertheless. Examples of totally hidden costs could be the organization's reputation, loss of customers, employee morale, etc. When measuring safety management performance, these costs, direct as well as hidden and totally hidden are an important statistic to record and monitor.

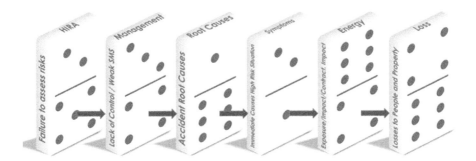

FIGURE 3.6 The accidental transfer of energy causes a loss in the form of injury, property damage or business interruption. (From McKinnon, Ron C. 2000. *The Cause, Effect and Control of Accidental Loss (With Accident Investigation Kit)*. Boca Raton: Taylor and Francis. With permission.)

SAFETY MANAGEMENT SYSTEM (SMS)

The loss causation analysis of an accident is of vital importance to the safety management profession. It calls for a different way of looking at, measuring, and controlling the prevention of occupational injuries, damage, and disease. The sequence clearly demonstrates the need for a structured SMS to identify the business risk and to institute ongoing management controls to prevent the sequence from occurring. The SMS prevents the accident domino effect by constantly identifying risks and setting up controls, balances, and checks that eliminate the root causes of accidents. This stabilizes the safety system control domino, preventing it from toppling and creating the domino effect.

SUMMARY

The accident sequence is triggered by a failure to adequately identify the hazards and assess the risks they pose, which in turn causes a lack of or inadequate control in the form of a weak or non-existent SMS.

The second phase of the accident sequence is a lack of or inadequate control of risks. This lack of control could be because of a weak or non-existent SMS, no safety system, no safety system standards and controls, or non-compliance with the standards.

The root causes of accidents are categorized as personal (human) and job (workplace) factors. They are the underlying reasons why high-risk behaviors are committed, and why high-risk conditions exist.

The fourth domino in the accident sequence depicts high-risk behavior and high-risk conditions, commonly and incorrectly referred to as the immediate causes of accidents. They are the immediate cause of the energy exchange.

The exposure, impact or contact, or exchange of energy is the part of the accident sequence that is most closely associated with the loss. The exposure, impact, or energy transfer is what injures, damages, pollutes, or interrupts the business process.

Traditionally, most SMS efforts focused on one specific type of loss – injury to people. Many situations that could have killed or injured workers never resulted in injury or death. A good SMS should also focus on and measure the non-loss-causing outcomes of undesired events. It should emphasize *pre-contact* control rather than *post-contact* control. It must direct its energy to prevent the event from occurring rather than depending on the consequence.

4 Defining Safety Management Performance

SAFETY MANAGEMENT

Safety management is the ongoing process of identifying the risks within the organization, identifying what actions need to be taken to reduce these risks, allocating safety authority, responsibility and accountability to the correct people and departments, setting standards of measurement for compliance to those standards, measuring against those standards, evaluating the conformance to these standards, and initiating corrective actions to rectify deviations.

PERFORMANCE

Performance is the act of performing, represented by initial and ongoing actions, and doing or carrying into execution the achievement or accomplishment. Performance is the execution of actions, deeds, or feats in order to fulfill a claim, promise, or request.

SAFETY MANAGEMENT PERFORMANCE

Safety management performance is the act of performing initial and ongoing safety management actions and carrying them into execution for the accomplishment of health and safety goals and objectives.

Safety management performance is operating or functioning with regard to the effectiveness of the health and safety management system (SMS). It is how effective the organization is at protecting the health and safety of its employees.

Safety management performance is a balance between the SMS achievements of the organization and the effectiveness of the risk mitigations that are implemented and maintained to enable the safety achievements.

SAFETY MANAGEMENT PERFORMANCE MEASUREMENTS

Safety management performance measurements are a set of methods and practices that organizations can use to track health and safety performance. They are leading and lagging data points that organizations can use to evaluate progress. In this process, an organization agrees to a specific set of performance indicators or objectives and then measures the progress toward those goals. These goals relate to key health and safety performance indicators.

Performance measurement is a process by which an organization monitors important aspects of its business. Safety management performance measurements are measurements of activities specifically undertaken to improve health and safety at a workplace or within an occupation.

Safety performance measurement is a process by which an organization monitors the processes, programs, and procedures within the SMS to see if they are achieving the goals and objectives set by the key performance indicators. Data is collected to reflect how its processes are working, and that information is used to drive an organization's decisions over time.

Performance measurement deals specifically with performance measures. These are the quantitative indicators put in place to track progress against strategy and objectives. Performance measurement asks, "How do we track the progress of the strategy we've put in place?" Good safety performance measures cover a wide variety of criteria, which can be grouped into two main areas:

- Leading measurements.
- Lagging measurements.

A COMPLEX CHALLENGE

In reference to measuring OSH (Occupational Safety and Health) performance, RoSPA (The Royal Society for the Prevention of Accidents) suggests that one of the reasons why occupational safety and health receives less board-level attention than other business priorities, is because of the difficulty in measuring effectiveness in responding to what is a complex, multi-dimensional challenge. In business, "what gets measured gets managed." However, in the field of occupational safety and health (OSH), it is far from clear whether "key players" share a clear view of performance and its measurement.

It also states that:

> Accident data should never be used as the sole measure of OSH performance and organizations should select a combination of indicators as OSH performance measures.

RoSPA Website (2023)

TYPES OF SAFETY MANAGEMENT PERFORMANCE MEASUREMENT INDICATORS

Safety management performance indicators can be either:

Quantitative – an indicator that can be counted or measured and is described numerically. For example, the number of safety inspections conducted, the number of employees trained in health and safety, and injury frequency rates.

Qualitative – an indicator that would describe or assess a quality or a behavior. For example, the standard of the safety training given, employee ratings of management commitment to achieving best practice in health and safety, and the thoroughness of accident investigations.

Defining Safety Management Performance

Leading Measurements

Leading safety management performance indicators are the processes, programs, procedures, and activities that are current and that take place continuously under the umbrella of the SMS. Lead indicators measure activities to prevent or reduce the probability and severity of an accident in the present or future. Examples include:

- The number of hazards identified and eliminated.
- The number of task risk assessments carried out.
- Numbers of employees trained in health and safety.
- Number of near miss incident reports received.
- Holding of safety committee meetings as per SMS standard.
- Number of safety inspections completed.
- Overall state of the SMS as verified by audit.

Leading indicators measure what occurs on an ongoing basis to reduce workplace risk before an accident occurs.

Lagging Measurements

Lagging safety performance measurements are indicators that follow accidental loss. Lag indicators measure outcomes after an accident and are effective measures of past results.

These include:

- Fatality rates.
- Injury rates.
- Severity rates.
- Property damage events.
- Costs of accidental losses.
- Days lost due to injuries or diseases.
- Loss of production.
- Downtime costs.

Lagging indicators are the result of accidental events and include accidents that damage property and disrupt normal production processes.

Soft Measures

Soft measures of safety performance are:

- Management's visible commitment to safety.
- Engaging and empowering workers in key decision areas for safety.
- Regular evaluation.
- Reduction of job hazards.
- Relevant and effective training.

These soft measures may be more difficult to measure than numbers of injuries and injury rates, but the soft measures almost certainly have a greater impact on performance. This may not be well recognized by many managers and safety professionals.

Hard Measurements

Hard controls and measurements are based more on the tangible, measurable aspects of the SMS. They measure the systems. Hard measurements are *objective* in nature, as it is easy to measure them. These include injury and loss statistics and scores derived from audits.

WHY MEASURE SAFETY MANAGEMENT PERFORMANCE?

Measuring safety management performance allows the review of an organization's health and safety performance over time, which may identify weaknesses in the system. Safety performance information also provides feedback on the effectiveness of controls in the form of programs, systems, activities, and procedures within the SMS.

Measuring safety management performance provides a forum for management review of the state of the SMS at the organization.

BENEFITS OF SAFETY MANAGEMENT PERFORMANCE MEASUREMENT

"If you don't know where you are going any road will lead you there." The measurement of safety management performance is vital to indicate to an organization where they are and where they should be going in terms of health and safety. Performance measurements will reveal the following:

- The status of the company's health and safety activities.
- If there has been an improvement in health and safety.
- The degree of accidental loss being experienced.
- How the organization compares with the industry standard.
- Whether all health and safety systems standards are up to date.
- Whether all legal requirements have been achieved.
- The strengths of the SMS.
- The weaknesses within the SMS.
- Opportunities for improvement of health and safety efforts.

SAFETY EFFORT AND EXPERIENCE (SEE)

Safety performance should be measured by taking into consideration the experience as well as the current efforts of the organization. This is the SEE (Safety Effort and Experience) concept, which considers the *safety effort* and *safety experience* of the organization. The *efforts* are the lead measurements, and the *lagging* ones are the experience measurements.

Defining Safety Management Performance

APPROPRIATE MIX OF MEASURES

An organization should use a mix of measures similar to looking ahead, to the side, and to the rear. To be successful, the organization must not just look back. The recommendation for effective application of safety measures is to put a major emphasis on looking ahead and using leading indicators. These leading indicators predict what will happen with the trailing or result measures. Trailing measures only look back. A balanced mix of leading and trailing measures would be recommended.

FOCUS

Instead of focusing on injuries alone, the organization should scan accidents for trends and focus on eliminating high-risk exposures and systematically increasing safe behaviors.

Ideally, the measurements should measure what drives safety within the organization.

MANAGEMENT REVIEW

A management review is the ideal opportunity for the health and safety activities of the organization to be scrutinized. These reviews are regularly held to discuss production and financial issues, and a separate review of the SMS is a preferred annual activity.

AGAINST STANDARDS AND OBJECTIVES

Safety management performance measurement must be against some form of standard, criteria, or benchmark so that the measurement becomes meaningful. Measurements can be internal and external.

Internal Measurements

Internal measurements are those measurements that are measured against internal company health and safety standards, targets, and objectives. These benchmarks could include:

- Target injury frequency rates (TIFR).
- Target injury severity rates (TISR).
- Days lost due to accidental injury or illness.
- Total cost of accidental losses.
- Hazards reported and rectified.
- Near miss incidents reported and rectified.
- Achievement of specific health and safety objectives.
- Employees trained in health and safety, etc.

External Measurements

External measurements are those that compare the organization with other organizations in the same industry classification. They can include:

- The industry nation average injury rate (incidence rates).
- The average workdays lost due to accidents.
- The energy exchange types experienced (event or exposure).
- Injuries per occupation.
- Part of body injuries.
- Injured worker characteristics (gender, age, ethnic origin, etc.).
- Injuries per state or region, etc.
- Achievement of national or international accreditation such as:
  - Occupational Safety and Health Administration Volunteer Protection Program (OSHA VPP).
  - International Organization for Standardization (ISO) ISO 45001:2018, (Occupational Health and Safety Management Systems).
  - Other national recognized accreditation for health and SMSs.

WHAT GETS MEASURED, GETS DONE

Many expect measurements of safety performance to determine if the organization is safe or unsafe. These concepts are too broad, hard to define, and have fuzzy measurements. What is meant by that is that some people's perception of *safety* would be different to another person's perception of *safety*. The expressions, *be safe* or *work safe*, are also wide open to interpretation and are not sound measurements of safety, nor is a *safe workplace*.

Safety Is a State

Based on the definitions of safety as a state of not being injured or harmed, the only way to gauge safety is if someone is injured or harmed. Then the person is no longer safe. The state would have changed. Unsafe means hazardous, perilous, or in danger. Therefore, the measure of safety is the consequence, which is an injury or illness. So, a safety measurement is a measurement of consequence, or a measure of injury, illness, or damage. The only way to measure a state is to conclude that if there is no injury or illness, the person is safe.

Too Simplistic

It would be far too simplistic to take the argument above and apply it to the workplace. If there were injuries in the workplace, it would be termed *unsafe*, and if there were no injuries, it would be termed *safe*. This type of gauge of safety is far too crude to be meaningful. What is meaningful, however, is the degree of workplace risk that the employees are exposed to and the efforts in place to mitigate those risks. That would be a far more accurate and meaningful measure of safety. A workplace that has reduced the risks to an acceptable level can be termed a safe workplace. An accepted level would be that the risks are as low as is reasonably practical.

IF IT CANNOT BE MEASURED, IT WILL NOT BE DONE

If the measurement of safety is not a substantial, meaningful gauge, then more than likely nothing will be done about it. What is needed are tangible, measurable, and doable metrics that can be quantified, calculated, and compared. Fuzzy concepts such as *safe* and *unsafe* do not fall into that category.

RISK PROFILE

A risk profile examines the nature and level of the threats faced by an organization, the likelihood of adverse effects occurring, the consequent level of disruption, and the costs associated with each type of risk. It also indicates the effectiveness of the controls in place to manage those risks.

The company risk profile indicates that the organization has identified hazards in the workplace, assessed the risks, and plotted a profile of risk in order of severity so that they can be mitigated on a priority basis. This risk profile is a positive measure of safety as it indicates that the sources of accidental injuries and illnesses have been identified and are being controlled, thus reducing the risk of workplace accidents. Risk profiles can be measured.

RISK FREE

No workplace can be without some degree of risk, so the concept of *risk-free* is also fuzzy and cannot be measured as it does not exist. The degree of risk reduction or actions to reduce risk can be measured and tracked.

ALARP

The main objective of any SMS is to drive the risks into the ALARP (As Low as Is Reasonably Practicable) zone. This reduction in risk is what is measured as well as the programs, processes, systems, and standards that lead to the ongoing reduction of these risks.

INPUT MEASUREMENTS (POSITIVE PERFORMANCE INDICATORS)

Input measures are sometimes referred to as positive performance indicators, as they are the efforts and actions that take place on an ongoing basis before accidents occur. Input measurements of health and safety are a combination of the actions, processes, and procedures that form part of the SMS and are aimed at reducing workplace risk.

PROCESS MEASUREMENTS

Process measures are used to monitor the SMS implementation (Figure 4.1). If SMS programs and procedures are not being implemented as intended, they cannot be expected to affect later measures of outcomes. Process measures monitor the amount and quality of a specific type of activity and output. The most common type of process measure is a counting system that keeps track of how much of something is being administered. For example, if an organization attempts to increase the number of near miss incidents reported, recording the number of submitted reports can be one measure of SMS delivery. Other types of process measures are more complex because they require more than a tallying system. Examples are the characteristics, behaviors, attitudes, opinions, and beliefs of individuals involved in an SMS.

Figure 4.1 shows a portion of a chart that measures the development and implementation of an SMS (in this example, 10 out of the total 73 SMS elements). It gives the element number, the element name, and the maximum score for implementation completed (20). The scores for each segment of the implementation are allocated on a (0–5) scale as follows:

Write – has the standard been written and approved? (0–5)
Revision – has the element been revised and approved and revised after 12 months or as needed? (0–5)
C or T – has a checklist been compiled and is it in use, and has training been presented concerning the contents of the standard? (0–5)
Implementation – to what degree has this SMS element been implemented throughout the organization? (0–5)

The scores for each element and the entire SMS implementation process are automatically updated by the chart.

OUTPUT MEASUREMENTS

Output health and safety measures are the outcomes or consequences of a failure or weakness in the SMS. They are lagging indicators generated by downgrading events that cause losses. They are also post-contact measures. These are measures of past performance.

SMS DEVELOPMENT PROGRESS CHART

Ele. No.	Health and Safety Management System – 73 Elements	Max	Write	Rev	C or T	Imp	Total	%
1.1	Building and floors	20	5	5	2	3	15	75%
1.2	Lighting: natural and artificial	20	5	5	2	3	15	75%
1.3	Ventilation: natural and artificial	20	5	5	2	3	15	75%
1.4	Occupational health facilities and service	20	5	5	1	2	13	65%
1.5	Pollution	20	5	5	3	3	16	80%
1.6	Aisles and storage demarcated	20	5	5	5	2	17	85%
1.7	Good stacking and storage practices	20	5	5	5	3	18	90%
1.8	Housekeeping	20	5	5	5	3	18	90%
1.9	Scrap and refuse bins removal	20	5	5	3	2	15	75%
1.10	Color coding: plant and pipelines	20	5	2	3	3	13	65%
	Section: 1 Premises and Housekeeping	200	50	47	31	27	155	78%

FIGURE 4.1 An example of a tracking system for the implementation of a health and safety management system (SMS).

CHARACTERISTICS OF GOOD SAFETY MANAGEMENT PERFORMANCE MEASUREMENT INDICATORS

For any health and safety performance measurement indicator to be effective, it is important that it is:

- Objective and easy to collect and measure.
- Understood and owned by the area, division, or workgroup whose performance is being measured.
- Relevant to the organization whose performance is being measured.
- Must provide immediate and reliable indications of the level of performance.
- Cost-efficient in terms of the equipment, personnel, and additional technology required to gather the information.

THE SAFETY RECORD

Many judge the state of health and safety at a workplace by the current *safety record*. In the past, this may have been an indicator, but records are traditionally based on lagging indicators, which don't appear until the record is broken by an accidental injury or disease. Safety records are based on injuries recorded or hours worked without injury. As long as the injury-free record remains intact, the organization is lulled into a false sense of security, believing that all is well. The absence of an injury may simply be fortuitous, and the record tells nothing of the safety efforts being made to reduce risk. The safety record is not a positive measurement of safety management performance as it could be inaccurate due to peer pressure and the safety fear factor.

PUBLIC PERCEPTION

The public's perception of an organization's safety status could also be influenced by the safety record quoted. Employee perceptions are accurate as they know the culture of the organization, and so-called records are often regarded by workers as shams. While the intent of the safety record may be well intended and may well be genuine, the achievement and maintenance of the record create opportunities for rule bending to avoid letting the team down by ruining the safety record.

DIFFERING RULES

One of the main downfalls of safety records is that the rules used to compile them differ. As will be discussed in later chapters, the definitions by which workplace injuries are defined are not always clear or universal. The application of the rules also differs from company to company.

SAFETY FEAR FACTOR

The safety fear factor is the fear (perceived or otherwise) that workers have of reporting personal injury as a result of a workplace accident. This fear factor could result in the non-reporting or underreporting of injuries, rendering the safety record invalid.

WORLD SAFETY RECORD

There have been many world records established based on injury-free periods at organizations.

According to the Guinness World Records:

> The greatest number of man/hours an industrial site has had a registered zero accidents, is 211,600,000 for Samsung Electronics Co., Ltd. Semiconductor Business Kiheung Site, Yongin City, South Korea between 4 Nov 1991 and 20 Aug 1998.
>
> **Guinness World Records (2023)**

The zero "accidents" should actually read, "zero injuries" (Editor).

The Austin American-Statesman (2022) reported in an article, *DuPont Must End its Safety Charade*:

> DuPont's STOP ignores all this. It tells companies that workers get hurt because they don't follow safety rules. They preach the way to fix safety problems is to monitor and discipline workers for "unsafe acts." The scheme also provides incentives – in the form of cash payments – for a "good" safety record. This only discourages workers from reporting injuries and hazards, which makes matters worse.
>
> **Austin American-Statesman (2022)**

SAFETY VERSES INJURY

Tallying injury-free periods is a traditional method of measuring safety management performance and tells little about the safety programs, systems, and processes in operation within the company. This means a loss must occur before a record is broken. Many organizations are so focused on injury-free periods and the safety record that they lose sight of the hazards and risks that exist in the workplace.

PARADIGM

A safety paradigm that exists is that if an organization is injury-free, then it must be a safe place to work and must care for the health and safety of its workers. Focusing entirely on a lagging measure of health and safety takes attention away from inputs intended to mitigate risks in an effort to reduce injuries and other forms of accidental loss.

SUMMARY

Safety performance measurement is a process by which an organization monitors the processes, programs, and procedures within the health and SMS to see if they are achieving the goals and objectives set by the key performance indicators.

Quantitative measures are indicators that can be counted or measured and are described numerically. Qualitative measures are an indicator that would describe or assess a quality or a behavior.

Defining Safety Management Performance

Measuring safety management performance allows the review of an organization's health and safety performance over time, which may identify weaknesses in the system.

"If you don't know where you are going any road will lead you there." The measurement of safety management performance is vital to indicate to an organization where they are and where they should be going in terms of health and safety. Safety performance should be measured by taking into consideration the experience as well as the current efforts of the organization.

Instead of focusing on injuries alone, the organization should scan accidents for trends and focus on eliminating high-risk exposures and systematically increasing safe behaviors. Ideally, the measurements should measure what drives safety within the organization.

Internal measurements are measured against internal company health and safety standards, targets, and objectives. External measurements compare the organization with other organizations in the same industry classification. If the measurement of safety is not a substantial, meaningful gauge, then more than likely nothing will be done about it.

Input measures are sometimes referred to as positive performance indicators, as they are the efforts and actions that take place on an ongoing basis before accidents occur. Process measures are used to monitor the SMS implementation. Output health and safety measures are the outcomes or consequences of a failure or weakness in the SMS.

Part II

Safety Management Performance Measurements

5 Pre-Contact, Contact, and Post-Contact Measurement

Management has three opportunities for safety management control. The same three opportunities exist for safety performance measurement.

PRE-CONTACT, CONTACT, AND POST-CONTACT CONTROL

There are three areas of safety control. They are the *pre-contact* phase, the *contact* phase, and the *post-contact* phase of the accident. Pre-contact activities are the interventions, programs, procedures, and processes that take place before an accident occurs. They are ongoing hazard identification and risk reduction programs and processes that occur according to the safety plan, objectives, and standards.

The post-contact stage occurs after the accident has occurred. Most safety efforts are aimed at the post-contact stage when a loss has already occurred. The accident has happened, and there is a sudden response to rectify the accident-causing problems. This is reactive and not pro-active safety. Instituting management control measures before the accident and subsequent losses occur, is control at the pre-contact stage, and is pro-active safety. Pre-contact management control of risk is facilitated by a structured health and safety management system (SMS) and its components.

PRE-CONTACT PHASE CONTROL

POSITIVE PERFORMANCE INDICATORS (PPIs)

Leading performance indicators, also referred to as positive performance indicators, allow measurement of activities specifically undertaken to improve the health and safety performance of an organization before accidents occur. Leading health and safety indicators are active steps that an organization can take to prevent future accidental events. Research has shown that measuring leading indicators is a great predictor of accidents. High-potential hazards and high potential near miss incidents are examples of accident precursors.

PRE-CONTACT CONTROL

Pre-contact control is the phase before the accident. It involves the systems, standards, and actions that are in place before any accidental contact with an energy source takes place. A few representative examples (programs, processes, and systems) of an SMS are:

- Hazard Identification and risk assessment.
  - Plant inspections.
  - Hazard reporting.
  - Risk assessment.
  - Risk control measures.
- Providing guards and barriers.
  - Setting standards for guarding.
  - Machine guarding surveys.
  - Regular inspection of guards.
- Good stacking and storage practices.
  - Good housekeeping standards.
  - Demarcation of aisles and work areas.
  - Correct storage and stacking.
- Safety awareness training.
  - Safety induction training.
  - First aid training.
  - General safety training.
- Critical task procedures.
  - Critical task identification.
  - Critical task procedures.
  - Training in the procedures.
- Work environment monitoring.
  - Noise zoning.
  - Lighting surveys.
  - Ventilation surveys.
- Ongoing safety inspections
  - Workplace inspections.
  - Equipment inspections.
  - Health and safety representative inspections.
- Buildings and floors.
  - Obvious unrepaired damage to buildings and floors.
  - Unsafe condition of floors.
  - Premises divided into areas and delegated to safety representatives.
- Other controls, programs, and processes within the SMS.

PRE-CONTACT PHASE – SAFETY PERFORMANCE MEASUREMENTS (LEADING INDICATORS)

Using the same representation sample of SMS elements, pre-contact safety performance measurements (monthly) would be:

- Hazard Identification and risk assessment.
  - Plant inspections – the number of inspections carried out per month.

Pre-Contact, Contact, and Post-Contact Measurement

- Hazard reporting – the number of hazards reported and rectified per month.
- Risk assessment – the number of risk assessments done per month.
- Risk control measures – the number and success of risk interventions.
- Providing guards and barriers.
  - Setting standards for guarding – is a standard written and is it available?
  - Machine guarding surveys – number of machine guarding surveys done per month.
  - Regular inspection of guards – inspections done per month.
- Good stacking and storage practices.
  - Good housekeeping standards – has a standard been written and is it applied?
  - Demarcation of aisles and work areas – the number of monthly inspections to measure against the standard.
  - Correct storage and stacking – the monthly degree of conformance to the standard for housekeeping and storage.
- Safety awareness training.
  - Safety induction training – the number of classes held monthly according to standard.
  - First aid training - the number of classes held monthly according to standard.
  - General safety training – the number of employees trained per month in other health and safety training classes.
- Critical task procedures.
  - Critical task identification – is the list updated?
  - Critical task procedures – are procedures used and updated as necessary (at least annually).
  - Training in the procedures – the number of operators trained in the procedures per month.
- Work environment monitoring.
  - Noise zoning – has noise zoning been done?
  - Lighting surveys – are illumination surveys done?
  - Ventilation surveys – are ventilation surveys done?
- Ongoing safety inspections.
  - Workplace inspections – the number and quality of inspections carried out per month.
  - Equipment inspections – the number and quality of inspections carried out per month.
  - Health and safety representative inspections – the number and quality of inspections carried out per month.
- Buildings and floors (measured monthly).
  - Obvious unrepaired damage to buildings and floors.
  - Unsafe condition of floors.
  - Premises divided into areas and delegated to safety representatives (on a plan or in writing).

Scoring Method

When measuring leading indicators, some form of scoring or ranking system is needed so that the activity can be quantified. A (Yes) or (No) scoring system is not accurate enough. Years of auditing experience have shown that a ranking of 0–5 for each criterion is easily understood, simple to use, and adequate for measurement purposes. The percentage representation is as follows:

1 represents 0%.
2 represents 20%.
3 represents 40%.
4 represents 60%.
5 represents 80%.
6 represents 100%.

Figure 5.1 shows an extract from an internal audit document showing the scoring system using the 0–5 scoring method. Irrespective of the measurement, whether it be pre-contact contact or a post-contact measurement, the result should be allocated a score or a number to quantify it. This simple scoring method measures the achievements of targets and ranks them as a percentage. A *yes* or *no* measurement is not definitive enough.

Minimum Standard Detail

Each element (which is one of approximately eighty which constitute a comprehensive SMS) of the SMS is broken down into minimum standards and minimum standard details, which can be measured on a 0–5 scale. If the element does not achieve 100%, it indicates there are weaknesses or failures in the SMS. This gives management opportunities to rectify weaknesses in the system before losses occur.

In Figure 5.1, the element of the SMS that is being measured is that of *Premises and Housekeeping*. In this example, this element has three minimum standard details. The total score for this element is fifteen points. Five points are allocated for

OBSERVATION GUIDE	SCORE (0=Low 5=High)						HAZARD	CONDITION COMMENTS	ACTION	
1.00 – Premises & Housekeeping	0	1	2	3	4	5	CLASS A-B-C		By	Date
1.1 - Buildings and Floors: Clean and in a good state and repair (15)										
Obvious unprepared damage to buildings										
Unsafe condition of floors										
Premises divided into areas and delegated to Safety Representatives/Supervisions (on plan or in writing)										

FIGURE 5.1 An example of the 0–5 ranking system.

obvious unrepaired damage to buildings and floors, five points are allocated to the *unsafe condition of floors and walkways*, and five points are allocated if the premises have been divided into areas, and *health and safety representatives have been appointed* in those areas to carry out monthly inspections.

MEASURING THE HAZARD BURDEN

The hazard burden is the range, type, distribution, and significance of the hazards within the organization and will determine the risks that need to be controlled. Various activities undertaken by an organization will create hazards, which will vary in type and significance.

Ideally, the hazard should be eliminated altogether by the introduction of a hierarchy of controls. If the hazard burden is reduced, this will result in a lower overall risk and a consequent reduction in injuries and ill health.

The hazard burden may increase as the organization takes on new activities or makes changes to existing ones. A change management element should be one element (program, process, or system) within the SMS.

Measuring the hazard burden will entail:

- Listing the hazards associated with production activities.
- Ranking the potential of the hazards (A, B, or C class hazards).
- Applying risk analysis techniques.
- Forecasting how the nature and significance of the hazards will vary over time.
- Recording the actions taken in eliminating or reducing the hazards.
- Determining what changes in the organization will have on the nature and significance of hazards.

CONTACT PHASE OF THE ACCIDENT

This is the phase in the accident sequence where the injury, damage, or other loss occurs. It is where accidental exposure, impact, or contact with an energy source and energy exchange take place. Previously, this was incorrectly termed the *accident*, but it is the segment of the accident where the energy is exchanged and the loss occurs. It is where the blade cuts into a finger, where the acid enters an eye, where a falling body strikes the ground or where harmful fumes are inhaled into the lungs.

CONTACT PHASE CONTROL

There is little opportunity for safety control at this stage in the accident sequence. Contact control is the provision of barriers, modifications to equipment and systems, and the supply of personal protective equipment (PPE). A control which is effective at the contact phase is the wearing of PPE. The PPE does not stop the flow of energy but reduces the consequences of the energy exchange. This contact control is deemed as the last resort and is only reverted to when all pre-contact efforts have been exhausted.

CONTACT PHASE – SAFETY PERFORMANCE MEASUREMENTS

The contact phase of the accident is where there is an exposure to a hazard, an impact with a substance or object, and an accidental exchange of energy which causes a loss in the form of illness, injury, property or equipment damage, or business interruption. Measuring the contact phase entails recording and measuring the types of energy exchange that were responsible for the losses. This is a measurement of the moment of impact, exposure, or energy transfer that occurred during the accident sequence.

Trends

This type of measurement does indicate trends and will identify agencies and agency parts responsible for accidental energy transfers. It will also quantify the transfer of energy types, previously incorrectly labeled *accident types*. While this measurement is not a pre-contact measurement, it does help to ascertain the types of energy exchanged and the agents involved so that trends can be compiled and pre-contact actions can be taken to prevent the recurrence of similar energy exchanges.

Types of Energy Exchange

There are numerous types and classifications of energy transfers and some of the main types are as follows:

- Struck against (running or bumping into).
- Struck by (hit by a moving object).
- Fall to lower level (either the body falls, or the object falls and hits the body).
- Fall on same level (slips, trips, and falls).
- Caught in (pinch and nip points).
- Caught on (snagged, hung).
- Caught between (crushed or amputated).
- Contact with (electricity, heat, cold, radiation, caustics, toxics, and noise).
- Overstress/overexertion/overload.

These energy transfer classifications would include exposures, impacts, and energy exchanges.

ACCIDENT AND NEAR MISS INCIDENT

Should there be an accidental exposure, impact, or energy flow and this inadvertently released energy does not contact a body or substance or is below the threshold limit of the body or substance, there will be no loss in the form of illness, injury, or damage, and the event is therefore termed a near miss incident. The energy flow missed the target. There was high potential for loss, but there was no loss. The person had a close call or a narrow escape. The actual energy exchange is what determines the difference between an accident (loss) and a near miss incident (no loss).

CASE STUDY

While removing the liners from a rotating mill, an employee left a 3-ft (900 cm) pry-bar lying in the bottom of the mill casing. To remove the liners, the nuts holding the liners in place are removed; employees then withdraw to a safe distance, and the mill is rotated, allowing the liners to fall to the bottom of the mill. On this occasion, the liner fell on a pry-bar which had been left in the mill, sending it flying through the air like a spear approximately 4 ft (1.2m) from the ground. Fortunately, it narrowly missed the supervisor as it flew past him at high speed. The potential of this near miss incident was ranked as extremely high, as was the probability of recurrence. Witnesses agree that under slightly different circumstances, the pry-bar could have hit the supervisor, causing serious injury or death.

All the factors of an accident were present as well as the flow of the energy of the pry bar, but there was no contact with the supervisor, and the bar narrowly missed him. If the bar had hit the supervisor, the energy of the pry bar would have been transferred (or exchanged), and this transfer of energy would have caused serious injury to the supervisor.

Recording and measuring the potential energy transfers of near miss incidents is an important aspect of contact phase measurement as well as pre-contact measurement.

POST-CONTACT PHASE

All actions that take place after an accidental loss has occurred are post-contact. They happen after the fact. Rendering first aid and evacuating injured workers are post-contact activities. Accident investigation is a post-contact activity. Near miss incident reporting and investigation is a pre-contact activity. This is why it is important to investigate high-potential near miss incidents. Many organizations make the mistake of trying to investigate all near miss incidents. This is a waste of effort. Reported near miss incidents should be risk-ranked as to their potential severity and probability of recurrence. Those with high potential and severity should be investigated.

Although accident investigation is an after-event activity, it is vital in order to set up pre-contact controls to prevent the recurrence of similar accidents in the future. It analyses the event, determines what went wrong, and paves the way for pre-contact controls to be implemented.

POST-CONTACT PHASE MEASUREMENT

Post-contact phase measurements are measurements of failure, as they can only be measured after an accident. After the inadvertent exchange of energy has taken place. This is a reactive measurement of safety management performance as it uses lagging indicators.

System Failure

Measuring post-contact results is an indication of a failure of the health and SMS. Some program, process, or activity has not functioned correctly to have caused a loss. This is the form of measurement that is typically used to gauge safety management performance, and some organizations have difficulty in understanding that it is not a reliable measure of safety. This form of measurement is most commonly used around the world, and a paradigm shift is required to encourage the use of leading as well as lagging indicators to provide a wholistic view of safety performance.

Accident-Based Metrics

These post-contact, lagging indicators are often referred to as "accident-based" metrics. They include injury-based measurements and the costs of accidental losses. Examples are:

- Fatality rates.
- Injury rates.
- Severity rates.
- Damage costs.
- Accident costs.
- Legal fines and penalties.

SUMMARY

Management has three opportunities for safety management control. The same three opportunities exist for safety performance measurement. The three opportunities are the pre-contact phase, the contact phase, and the post-contact phase of the accident. Pre-contact activities are the interventions, programs, procedures, and processes that take place before an accident occurs. They are ongoing hazard identification and risk reduction processes that occur according to the safety plan, objectives, and standards.

Contact measurements are measurements of the energy exchange type and the type of agency and agency part involved in the loss.

All actions that take place after an accidental loss has occurred are post-contact. They happen after the fact. Rendering first aid and evacuating injured workers are post-contact activities. Accident investigation is a post-contact activity. Post-contact measurements include injury rates, severity rates, and accident costs. Organizations should make use of leading indicators as well as lagging indicators to provide a wholistic view of safety performance.

Part III

Measuring Lagging Indicators of Safety Management Performance

6 Lagging Indicators of Safety Performance

Injuries, Illness, and Diseases

According to the publication, *Safety in Numbers,* International Labor Organization (ILO) (2003b):

> Decent work is safe work, but we are a long way from achieving that goal. Every year some two million men and women lose their lives through accidents and diseases linked to their work. In addition, workers suffer 270 million occupational accidents and 160 million occupational diseases each year - these are conservative estimates.

p. 1

INTRODUCTION

The only seemingly visible evidence of safety failure or success seems to be the fatality and injury numbers. Organizations, departments, and institutions are established to record, tabulate, and produce fatality and injury numbers on a full-time basis. These figures are used to gauge the safety of a country, state, industry type, workplace, or company.

Fatality and injury rates are seemingly the only evidence of safety that is produced and published on a regular basis. These numbers are what legal bodies, management, and the public associate with safety.

Injury or fatality-free records are also produced and are quoted as positive safety statistics. Fatality and injury tallies are all accident-related numbers.

MEASURES OF FAILURE

Other than the injury- or fatality-free records that are publicized, most statistics are lagging indicators as they are generated because of accidents. The statistics derived from undesired events such as accidents cannot be termed indicators of safety success but rather indicators of safety failure. An accident can be regarded as a failure of some component within the safety management system (SMS), and therefore the resultant illness or injury is a result of failure. Decreases in injury rates should not, therefore, necessarily be viewed as improvements in safety but rather as a decrease in the degree of safety failure.

MEASURES OF CONSEQUENCE

Lagging indicators of safety performance are measures of consequence – the consequence of a weakness or failure in the SMS. Many would like to believe that these were failures of the employees alone, but management is ultimately responsible for the safety of employees, and if the systems designed to protect them from hazards within the workplace fail, then it is a management failure.

REAR-VIEW MIRROR

Lagging indicators of safety management performance tell management where they have been. Like looking in a rear-view mirror. Management wants to know where they are going and not where they have been. Leading indicators tell management where they are going.

AFTER THE EVENT

Lagging indicators are obtained after the event has occurred. An accident must occur and cause illness, injury, or property damage before a statistic can be compiled. If there are no injuries caused by accidents, the organization tallies an injury-free period as a *safety record*. The lag indicator centers around the occurrence or nonoccurrence of an event, such as an accident.

CONSEQUENCES

The consequences of an accident vary from event to event. Some cause injury or occupational illness and disease. Some accidents only cause property damage and business interruption. Since lagging indicators are consequence based, all accidental consequences should be recorded and measured, not just injury or illness outcomes.

OUTCOMES

The reason for recording all outcomes of undesired events and not just injuries or illnesses is that under different circumstances, the outcomes could have been different. Since these outcomes are neither predictable nor controllable, the events and all their outcomes should form part of the lagging indicator statistics. These outcomes would include illnesses, occupational diseases or disorders, injuries, fatalities, property and equipment damage, and business interruptions.

ADVANTAGES OF INJURY AND ILLNESS DATA

Injury rates take on more meaning for an employer when the injury and illness experience of the organization is compared with that of other employers doing similar work with workforces of a similar size. This information permits detailed comparisons by industry category and size of workforce. It also enables management to benchmark against competitors or the best-in-class in their industry. The rates also indicate if the organization's injury rate is improving or worsening over the years. It also informs whether or not the organization's injury rates are within the legal range for the type of industry.

DISADVANTAGES

Lagging indicators reflect injury and illness histories that cannot be changed. They indicate failures in the safety management process and are a record of the consequences of these failures. The data received is not always accurate, is subject to manipulation, and does not give management a clear picture of the state of safety within the organization.

MEASUREMENT OF CONTROL

Leading indicators show management where they are going. They are predictive rather than reactive. Leading indicators do not need a downgrading event to occur first. They offer advanced opportunities to avoid accidental losses.

Although there is no guarantee that monitoring leading indicators will eliminate accidents entirely, they offer the opportunity to identify them and reduce the probability of the event occurring.

GUIDELINES TO RECORDING INJURY EXPERIENCE

There are many guidelines, definitions, and descriptions of terms and criteria used in calculating lagging indicators, and these vary from country to country. For many years, the 1954 and 1967 revised versions of the American National Standards Institute's (ANSI) ANSI Z.16.1 *American National Standard Method of Recording and Measuring Work Injury Experience* provided guidelines for the calculation of injury and fatality rates. This was perhaps the most comprehensive and detailed standard for recording and measuring workplace injury experience but was not always followed by all workplaces. Many countries have their own version of reporting and recording workplace injuries as prescribed by local health and safety legislation. The following descriptions are generally used in the calculation of injury statistics.

EMPLOYMENT

Normal employment means all work or activity performed in carrying out an assignment or request of the employer, including incidental and related activities not specifically covered by the assignment or request, as well as any voluntary work or activity undertaken while on duty with the intent of benefiting the employer. It includes any activities undertaken while on duty with the consent of the employer.

OCCUPATIONAL INJURY

An occupational injury (which includes illness and occupational disease) is any injury suffered by a person that arises out of and during his or her employment.

OCCUPATIONAL FATALITY

An occupational death is any fatality resulting from a work injury, regardless of the time intervening between injury and death. In fatal cases where death might have resulted either from an illness or from an accident following the illness, the case may be considered a work injury only if it is the opinion of the attending physician engaged or authorized by the employer that the illness arose out of, or was aggravated by, the victim's work. For example, a worker died after falling from a scaffold he had just climbed. There was evidence of a heart attack, and if it was the opinion of the doctor who handled the case that climbing the ladder had contributed to the heart attack, the case could be considered a work fatality.

OCCUPATIONAL FATALITY RATES

The fatality rate is the number of fatalities per 100,000 employees. To calculate the fatality rate, divide the number of fatalities by the total number of hours worked by the organization or industry type and multiply by 200,000,000. The number of fatalities experienced and the number of employees could be for a company, state, or country.

The formula is:

$$\text{Fatality Rate} = \frac{\text{Fatalities}}{\text{Total employee workhours}} \times 200{,}000{,}000$$

According to the US Bureau of Labor Statistics (BLS) (2023), there were 4,764 fatal work injuries recorded in the United States in 2020, a 10.7% decrease from 5,333 in 2019. The fatal work injury rate was 3.4 fatalities per 100,000 full-time equivalent (FTE) workers, down from 3.5 per 100,000 FTE in 2019 (BLS Website).

HISTORIC MEASURE

Workplace fatalities have been recorded by industries and mines for years. It is perhaps the most obvious consequence of an accident and the worst workplace loss that can be experienced. Major construction projects in the past resulted in multiple workplace fatalities, which represented only the tip of the iceberg.

GOLDEN GATE BRIDGE

During the construction of the Golden Gate Bridge, eleven men died. The first fatality was on October 21, 1936. Then, on February 17, ten men lost their lives when a section of scaffold fell through the safety net, which did save the lives of 19 men who became known as the "Halfway-to-Hell Club."

In the 1930s, a rule of thumb on high-steel bridge construction projects was to expect one fatality for every $1 million in cost. The construction safety record for the $35 million Golden Gate Bridge was 11 construction workers who died. By contrast, 28 laborers died building the neighboring San Francisco-Oakland Bay Bridge, which opened six months prior.

Hoover Dam

The official number of fatalities involved in building Hoover Dam is 96. These were men who died at the dam site from such causes as drowning, blasting, falling rocks or slides, falls from the canyon walls, being struck by heavy equipment, truck accidents, etc. Not included in the official number of fatalities were about 40 deaths that were recorded as pneumonia. Workers alleged that this diagnosis was a cover for deaths from carbon monoxide poisoning brought on by the use of gasoline-fueled vehicles in the diversion tunnels.

High-rise Construction Fatalities

The Eiffel Tower kept its construction worker death toll down to one worker, with much credit going to the extensive use of guard rails and safety screens. The Empire State building had five deaths among its 3,400 workers. The 1970 World Trade Center construction project recorded 60 construction worker deaths. The Sears Tower recorded five worker deaths in two incidents, and Las Vegas's City Center project resulted in the deaths of six construction workers.

Deadliest Project

The Panama Canal is by far one of the deadliest construction projects, with an official total of 5,609 deaths during the US construction era. On the opposite end of the spectrum, the Chrysler Building had 3,000 workers and zero construction worker deaths.

World Cup 2022 Qatar

According to a report from *The Guardian*,

> Construction on venues for the 2022 FIFA World Cup in Qatar began immediately after the small Middle Eastern country was awarded the event in December 2010. Since then, more than 6,500 migrant construction workers have died.

> While death records are not categorized by occupation or place of work, it is likely many workers who have died were employed on these World Cup infrastructure projects as a significant proportion of the migrant workers who have died since 2011, were only in the country because Qatar won the right to host the World Cup.

Website (2022)

FATALITY-FREE SHIFTS

In hazardous, labor-intensive mines, projects and industries, fatality-free shifts are a common measurement of safety management performance.
According to the E&MJ Engineering and Mining Journal (2023),

> On January 26, 2020, Sibanye-Stillwater's South African gold operations achieved 10 million fatality-free shifts. This is a significant milestone, which has never been

achieved in the history of these gold operations, nor in the history of the South African deep-level gold mining industry, according to the company.

Website (2023)

LOST TIME INJURY

The lost-time injury is perhaps the most widely used measure of safety performance. Most countries use the lost-time injury rate as a prime safety metric. This classification of injury distinguishes it from minor injuries in that the injury is so severe that the injured person loses time away from work. The time lost must be a complete work shift or more other than the shift on which the injury occurred. Minor injuries and first aid cases, however, allow the injured person to be treated and to return to normal employment.

The definition of a lost time injury is important to ensure the same criteria are used when benchmarking or for other comparative reasons. Universally, there is no one definition of a lost time injury. The rules differ from country to country and from organization to organization.

TEMPORARY TOTAL DISABILITY (LOST TIME INJURY)

A lost-time injury (or disabling injury) results in temporary total disability. This is any injury that does not result in death or permanent impairment but renders the injured person unable to carry on his or her normal activities of his or her employment (the job he or she normally does) during the entire time interval corresponding to the hours of his or her regular shift on any one or more days (including Fridays, days off, or plant shutdowns) subsequent to the date of the injury. This also includes all fractures and bone damage, including those that do not result in permanent impairment or restriction of the normal function of the injured member or result in no lost time.

SHIFT LOST

An accepted rule when defining lost-time injuries is that the injured person must lose a shift or longer because of the injury, excluding the shift on which he or she was injured. All workplace injuries cause time loss, but this classification means that an entire work shift must be lost because of the injury. This means the injured person could not return to their normal duties for a shift or longer. Transferring the injured person or putting them on light duty does not distract from the fact that they were temporarily disabled and unable to continue with their normal work assignment.

REGULAR JOB

Many definitions call for the injured worker to return to his or her *regular established job* after the injury, and this would not classify the injury as a lost-time injury. Some interpretations of the requirements of a lost-time injury are that the injured person is unable to perform his or her *normal work or role* on the calendar day following

Lagging Indicators: Injuries, Illness, and Diseases 63

the day of the injury. Establishing *normal work*, a *normal role*, or an *established job* opens the definition of a lost-time injury to manipulation. A *routine function*, according to the Occupational Safety and Health Administration (OSHA), is one the employee would normally perform at least once a week.

THE ACID TEST

The definition of a lost time injury was modified to require the injured person to return to the *task he or she was doing at the time of the accident*. If the injured person was loading boxes onto a rack at the time of the accident, could they return to loading the same boxes on the same rack after the injury without losing a work shift?

Giving an injured person a less demanding job after the accident because it was part of their normal employment defeats the objective of injury recording.

INJURY FREQUENCY RATES

An injury frequency rate is the number of injuries (including occupational illnesses and diseases) experienced per 500 employees working for a year. Five hundred workers work approximately one million workhours a year, so the frequency rate equates to injuries per million workhours of exposure.

UNIVERSALLY USED

Injury frequency rates have come to be recognized as an accepted indicator of safety management performance. Many are familiar with the term injury rate, and by using this injury rate, comparisons can be made with other industries of a similar type. The formula converts the actual number of injuries experienced to injuries experienced per 1,000,000 workhours.

HISTORY

When injury rates were first introduced, the larger organizations employed in the region of 500 workers, so the injury rate was centered around one million workhours (approximately the equivalent of 500 employees working 40 hours per week, 50 weeks per year), and we used that number for comparisons between these similar-sized organizations.

COMPARISON

Industry classifications, such as the agriculture industry, can compare themselves with other agricultural industries by using injury frequency rates. National average injury rates for the industry classification can be compiled, and the organization can gauge its performance against this national average rate. These rates can also be used for benchmarking purposes both internally and externally, as well as to gauge the injury performance of the organization.

NATIONAL FIGURES

The US Bureau of Labor Statistics (BLS) (2023) reports that in the private industry, employers reported 2.6 million nonfatal workplace injuries and illnesses in 2021, a decrease of 1.8% from 2020. In 2021, the incidence rate of total recordable cases in private industry was 2.7 cases per 100 FTE workers, unchanged from 2020. The decline in injury and illness cases was due to a drop in illness cases, with private industry employers reporting 365,200 nonfatal illnesses in 2021, down from 544,600 in 2020, a drop of 32.9% (BLS Website).

LEGAL RATING

Many legal health and safety entities use injury rates to identify workplaces that are above the industry average, which in some cases prompts a site visit by inspectors.

LOST TIME INJURY FREQUENCY RATE (LTIFR) OR DISABLING INJURY FREQUENCY RATE (DIFR)

The lost time injury frequency rate (LTIFR) is the number of lost time or disabling injuries experienced for every 1,000,000 workhours of exposure. The period during which the injuries occurred and the exposure period must be the same. This is normally achieved by plotting a 12-month rolling average where 12 months of workhour exposure and injuries are used (Figure 6.1).

Formula

The formula for the LTIFR is:

$$\text{LTIFR} = \frac{\text{Lost time injuries}}{\text{Workhours}} \times 1,000,000$$

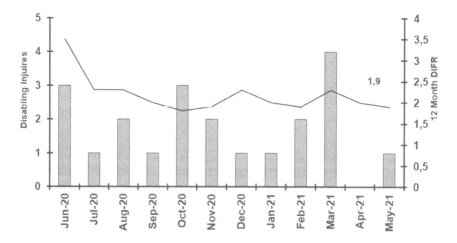

FIGURE 6.1 The graph shows the actual number of disabling or lost-time injuries per month and the rolling 12-month average LTIFR or disabling injury frequency rate (DIFR), which is calculated per million workhours.

Lagging Indicators: Injuries, Illness, and Diseases

The DIFR formula is:

$$\text{DIFR} = \frac{\text{Disabling injuries}}{\text{Workhours}} \times 1,000,000$$

INJURY INCIDENCE RATES

Injury incidence rates use the same criteria for the classification of injury but equate the rate per 200,000 workhours, which equates to 100 employees working for a year. The 200,000 hours in the formula represent the equivalent of 100 employees working 40 hours per week, 50 weeks per year, and provide the standard base for the incidence rate. The resultant figure is the percentage of employees suffering injuries per year, which is an easier number to comprehend.

LOST TIME INJURY INCIDENCE RATES (LTIIR) OR DISABLING INJURY INCIDENCE RATE (DIIR)

The lost time injury incidence rate is the number of lost time injuries experienced per 200,000 workhours, or the percentage of employees suffering lost time injuries during the measurement period (Figure 6.2).

The formula for the lost time injury incidence rate (LTIIR) and the disabling injury incidence rate (DIIR) is:

$$\text{LTIIR} = \frac{\text{Lost time injuries}}{\text{Workhours}} \times 200,000$$

$$\text{DIIR} = \frac{\text{Disabling injuries}}{\text{Workhours}} \times 200,000$$

The same formula would be used for all injury *Incidence Rates*.

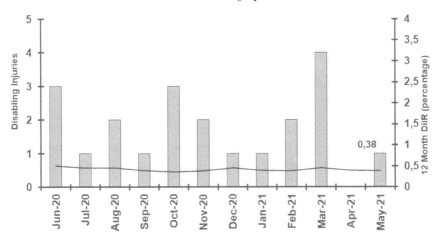

FIGURE 6.2 The graph shows the number of lost time or disabling injuries per month and the 12-month rolling average of lost time or DIIR per two-hundred thousand workhours.

TOTAL INJURY RATE

Total injury and illness rates can be calculated as frequency or incidence rates. This rate would include all injuries experienced at the organization within the recording period. First aid cases as well as medical treatment injuries, lost time, and all other injuries, illnesses, or muscular skeletal disorders would be included for calculation purposes. The resultant figure would give the percentage of employees experiencing any form of injury or illness during the measurement period. The total injury incidence rate (TIIR) is expressed per 200,000 workhours. The formula for the TIIR is:

$$\text{TIIR} = \frac{\text{Total injuries and illnesses}}{\text{Workhours}} \times 200,000$$

The Total Injury Frequency Rate (TIFR) is expressed per million workhours of exposure.

$$\text{TIFR} = \frac{\text{Total injuries and illnesses}}{\text{Workhours}} \times 1,000,000$$

THE TOTAL RECORDABLE DISEASE FREQUENCY RATE

The total recordable disease frequency rate (TRDFR) is the number of work-related recordable diseases per 1,000,000 workhours of exposure. The TRDFR is calculated for the recording period as:

$$\text{TRDFR} = \frac{\text{Total recordable diseases}}{\text{Workhours}} \times 1,000,000$$

OSHA RECORDABLE INJURY OR ILLNESS

An Occupational Safety and Health Administration (OSHA) recordable injury is a term for injuries and illnesses that must be reported to OSHA on Form 300 (log of work-related injuries and illnesses). It includes a work-related injury or illness that results in any of the following:

- Medical treatment beyond first aid.
- Loss of consciousness.
- One or more days away from work following the date of the incident.
- Restricted work or transfer to another job.
- Any significant injury or illness diagnosed by a physician or other licensed health care professional.
- Any work-related fatality.

DART

Days away, restricted, or transferred (DART) is a safety metric used by the Occupational Safety and Health Administration (OSHA) to show how many workplace injuries and illnesses caused the affected employees to remain away from

Lagging Indicators: Injuries, Illness, and Diseases   67

work, restricted their work activities or resulted in a transfer to another job as they were unable to do their usual job.

The DART rate defines the number of recordable injuries and illnesses per 100 full-time employees, which resulted in lost workdays, restricted workdays, or job transfer due to workplace injuries or illnesses. It can be calculated using the following formula:

$$\text{DART} = \frac{\text{Recordable injuries (days away, restricted, transferred)}}{\text{Workhours}} \times 200,000$$

The DART rate is designed to track any OSHA-recordable work-related injury or illness that results in time away from work, restricted job roles, or an employee's permanent transfer to a new position.

THE TOTAL CASE INCIDENT RATE

The *total case incidence rate*, also known as the TCIR and TRIR (total recordable incident incidence rate), is defined as the number of all work-related injuries per 100 full-time workers. The TCIR is a figure that represents the number of work-related injuries per 100 full-time workers over the course of a year. The calculation is based on the number of mandatory reported OSHA-recordable injuries and illnesses. Because of this, TCIR is also known as the OSHA incident rate. The formula is:

$$\text{TCIR} = \frac{\text{Total OSHA} - \text{recordable injuries and illnesses}}{\text{Workhours}} \times 200,000$$

MINE SAFETY AND HEALTH ADMINISTRATION (MSHA) REPORTABLE INJURY

The term *reportable injury* as defined by MSHA, includes all incidents that require medical treatment, or result in death, or loss of consciousness, or inability to perform all job duties on any workday after the injury, or temporary assignment or transfer to another job. Injuries involving "first aid only" are not reportable. First-aid only is defined as one-time treatment and subsequent observation of minor scratches, cuts, burns, splinters, and so forth, which do not ordinarily require medical care, even if it was provided by a physician or a registered professional person.

MSHA INCIDENCE RATES

The US Mine Safety and Health Administration (MSHA) recordable injuries are divided into 3 categories:

- Fatal injuries are work-related injuries resulting in the death of employees on active mine property.
- Nonfatal days lost (NFDL) cases are occupational injuries that result in the loss of one or more days from the employee's scheduled work or days of limited or restricted activity while at work.

- No days lost (NDL) cases are injuries requiring only medical treatment – beyond first aid.

Incidence rates, as defined by MSHA, are the number of injuries in a category, times 200,000 divided by the number of employee-hours worked:

$$IR = \frac{\text{Injuries in a category}}{\text{Workhours}} \times 200,000$$

RIDDOR (REPORTING OF INJURIES, DISEASES AND DANGEROUS OCCURRENCES)

As required by UK health and safety law, RIDDOR (The Reporting of Injuries, Diseases and Dangerous Occurrences Regulations) puts duties on employers, the self-employed, and people in control of work premises (the Responsible Person) to report certain serious workplace accidents, occupational diseases, and specified dangerous occurrences (near misses). Several types of dangerous occurrences require reporting in circumstances where the incident has the potential to cause injury or death.

The RIDDOR incidence rate for injuries that cause workers to lose more than 3 days of work, calculated per 100,000 workers, is:

$$\text{Incidence Rate 3-Day Absence} = \frac{3-\text{Day absence injuries (PA)}}{\text{Number of employees (PA)}} \times 100,000$$

The RIDDOR annual incidence rate (AIR) formula is calculated as follows:

$$\text{AIR} = \frac{\text{Fatalities} + \text{RIDDOR specified injuries} + \text{RIDDOR injuries}}{\text{Average number of employees in 12 month period}} \times 100,000$$

INJURY SEVERITY RATES

Disabling (lost time) injury severity rates (DISR) (Figure 6.3) measure lagging safety performance in terms of the workdays lost due to work-related injuries. The time lost is dependent on the severity of the injury. The injury severity rate is the number of shifts or workdays lost as a result of an injury. The severity rate can be expressed as shifts lost per one million or 200,000 workhours. The injury severity rate formula is:

$$\text{DISR} = \frac{\text{Work days or shifts lost}}{\text{Workhours}} \times 200,000$$

WORKDAYS LOST DUE TO WORK-RELATED ACCIDENTS

In the US, 65 million workdays were lost due to workplace injuries in 2020. The average number of days lost per individual injury varies from 8 to 11 workdays. The 1954 and 1967 revised versions of the American National Standards Institute (ANSI)

Lagging Indicators: Injuries, Illness, and Diseases

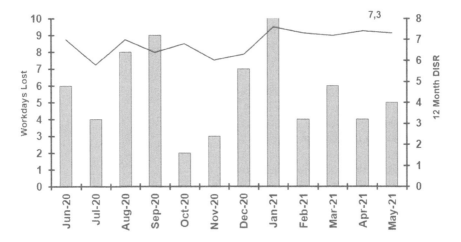

FIGURE 6.3 The graph shows the number of workdays lost per month as a result of injuries and the 12-month rolling average disabling injury severity rate of workdays lost per million workhours.

ANSI Z.16.1 *American National Standard Method of Recording and Measuring Work Injury Experience* give the workdays to be charged per fatal and permanent total disability accidents as 6,000 workdays.

AVERAGE DAYS CHARGED PER DISABLING INJURY

The average number of days lost per disabling injury is obtained by dividing the total days charged by the total number of disabling injuries.

$$\text{Average days charged per disabling injury} = \frac{\text{Total days charged}}{\text{Total number of disabling injuries}}$$

DISABLING INJURY INDEX

A disabling injury index (DII) is an index computed by multiplying the DIFR (injuries per million workhours) by the disabling injury severity rate (DISR) (shifts lost per million workhours) and dividing the product by 1,000. This measure reflects both frequency and severity, yielding a combined index of total disabling injury. The formula for the DII is:

$$\text{DII} = \frac{\text{Disabling injury frequency rate} \times \text{Disabling injury severity rate}}{1000}$$

FIRST AID DEFINED

According to OSHA, first aid refers to medical attention that is usually administered immediately after the injury occurs and at the location where it occurred. It often consists of a one-time, short-term treatment and requires little technology or training to administer. First aid can include:

- Cleaning minor cuts, scrapes, or scratches.
- Treating a minor burn.
- Applying bandages and dressings.
- Use of non-prescription medicine.
- Draining blisters.
- Removing debris from the eyes.
- Massage.
- Drinking fluids to relieve heat stress.

Workplaces are required to provide basic first aid measures and, in some instances, a first aid room where first aid can be administered. Other legal requirements are having trained first aides on all work shifts or, in some cases, a full-time first aid attendant.

MEDICAL TREATMENT DEFINED

Medical treatment is defined as occurring when an injury or disease requires a higher degree of patient management to ensure a full recovery. Medical treatment beyond first aid is a criterion that determines if a work-related injury or illness in the US is OSHA-recordable. Any work-related incident where the involved parties received medical treatment other than first aid is considered OSHA-recordable. Examples of medical treatment are:

- Suturing of wounds.
- Treatment of fractures.
- Treatment of bruises by drainage of blood.
- Treatment of second- and third-degree burns.
- Providing prescription drugs or non-prescription drugs.

FIRST AID INJURY RATE

First aid injury rates are the number of first aid cases recorded over a specified period. These could include treatments as a result of injury due to workplace accidents or visits to the first aid room for non-injury-related treatments such as headaches, dust in the eye, or insect bite treatment. The metric could be the total number of visits to the first aid room or visits per employee on average over a specified period. First aid treatments for work-related injuries would be included in the minor injury incidence rate.

MINOR INJURY RATE

The minor injury incidence rate (MIIR) is the total number of minor injuries, including first aid and medically attended injuries but excluding disabling or fatal injuries, per 100 employees per year. The formula is:

$$\text{MIIR} = \frac{\text{Minor injuries}}{\text{Workhours}} \times 200,000$$

The minor injury rate can also be expressed as a minor injury frequency rate by substituting 200,000 workhours with one million workhours.

TARGET INJURY FREQUENCY RATE (TIFR)

A target injury frequency rate (TIFR) is a goal set to be achieved based on a disabling injury frequency rate. The company could set a goal for the frequency rate to be lower than the national average for the industry or for an improvement in the current rate. A common target is a TIFR of five or a target injury incident rate (TIIR) of one, as this equates to a 1% incidence rate, equivalent to 1% of the workforce experiencing disabling injuries per year.

ANALYSIS OF HIGH-RISK BEHAVIOR AND HIGH-RISK WORKPLACE CONDITIONS

A lagging indicator could include an analysis of high-risk behavior and high-risk conditions. This analysis would indicate which high-risk behaviors and conditions are responsible for the majority of injuries. Trends can be established so that preventative measures can be put in place, reinforced, or enforced.

EXPOSURE, IMPACT OR ENERGY EXCHANGE, TERMINOLOGY

For many years, occupational safety nomenclature termed the type of *energy exchange* as the *accident type*. The correct term is type *of energy exchange*, or *nature of energy transfer or exposure*. For example, *contact with electrical current* describes the unintentional flow of energy that caused the injury. It describes contact with a source of energy greater than the threshold resistance of the body or substance, which results in a loss, or in this case, injury.

ENERGY TRANSFER (EXCHANGE) TYPES

Another form of lagging indicator of safety performance is the recording and classification of the energy transfer types that were responsible for the injuries or illnesses. This analysis provides important information as it indicates where weaknesses in the system lie that resulted in these energy exchanges. The analysis helps put remedial measures in place to prevent future energy transfers. Each injury or illness, including musculoskeletal disorders, is caused by exposure, impact, or energy transfer. Analyzing these would help identify the immediate and root causes of these contacts.

Common classifications of energy transfers and examples are:

- *Struck by an Object or Equipment* – While walking from the office to the warehouse, the injured person was struck by a moving truck and suffered a broken leg.
- *Struck Against Object or Equipment* – While walking down the aisle between the boiler and the work area, a worker misjudged the corner and

bumped into the side of the storage rack, cutting his right leg. His leg struck against the storage rack.
- *Caught in or Compressed by Object or Equipment* – Trying to take a short cut through the factory, the worker moved behind a reversing forklift truck, which crushed him against the wall, causing serious internal injuries. He was caught between the wall and the vehicle.
- *Fall to Lower Level* – While painting a wall, a worker slipped from the handrail he was leaning against and fell to the floor level below. He was injured as a result of his fall to a different level.
- *Fall on the Same Level* – While changing a light fitting, the maintenance man lost his balance and fell off the ladder, landing on the floor. He fell on the same level.
- Slip, trip *with a fall* – While walking through the processing area, the employee was reading messages on her cell phone and did not see the notice warning of a wet floor ahead. She slipped and fell onto the floor, injuring her elbow. The injury was caused by a fall on the same level.
- *Slips, Trips without Fall* – Because of poor housekeeping, an employee tripped over an obstruction, almost fell, and twisted her back. Although she did not fall, the trip caused her to jerk her body, which injured her back.
- *Overexertion in Lifting or Lowering* – Failing to heed to the basic rules of lifting heavy objects, the employee tried to move a large box from the back of the vehicle and hurt her back. The weight was too heavy and awkward for her, and she overexerted her back.
- *Repetitive Motion Involving Microtasks* – Typing on a keyboard for long periods, without frequent breaks, caused a worker to suffer from tendonitis as a result of the repetitive motion. Tendonitis is also sometimes referred to as trigger finger, or tennis elbow. Other repetitive strain injuries (RSI) are carpal tunnel syndrome and bursitis, which cause weakening of the movement of a limb and which can be very painful.
- *Roadway Incidents Involving Motorized Land Vehicles* – A truck driver was injured when his vehicle was involved in a multiple vehicle pileup on the highway. The pileup was due to poor visibility caused by mist. This category does not include injuries caused by water vessels. This energy exchange type was the cause of most work-related fatalities in the US during 2019.
- *Intentional Injury by Other Person* – Tempers flared among employees at a foundry as a result of a spoilt batch of aluminum castings. One employee had operated the crane before being instructed to do so, resulting in the batch of castings being ruined. The foreman was so angry, he pulled the crane operator out of the cab and struck him with his fist.
- *Injury by Person – Unintentional or Intent Unknown* – A construction worker accidentally and without intent dropped a piece of lumber onto a fellow worker below. The timber struck the employee on the shoulder, causing

an abrasion. The injury was caused by the unintentional action of a fellow employee.
- *Animals and Insects* – Animals and insects can cause serious injury to workers on the job. In one case, a farm worker was attacked by a bull and suffered serious injuries. In another case, while installing an outside light, an electrician disturbed a wasps' nest and was stung by the angry wasps.
- *Fires and Explosions* – Fires and explosions cause serious and fatal injuries to employees. Fires and explosions were the sixth-highest cause of work fatalities in the US during the year 2019.
- *Exposure to Harmful Substances or Environments* – Asbestos-laden atmospheres, areas where toxic gases are present, and other environments that contain harmful substances, can lead to serious harm or fatal injuries to employees. Some toxic environments can be entered into for brief periods without harm. Some cannot be entered into at all without the risk of being overcome and injured. Smoke, fumes, and mist are other examples of hazardous environments. Noise zones also fall into this category, as do areas where excessive radiation is present.

TOP THREE CAUSES

The top three causes (energy transfer types) that account for more than 75% of all nonfatal injuries and illnesses involving days away from work in 2020 were:

- Exposure to harmful substances or environments.
- Overexertion and bodily reactions.
- Slips, trips, and falls.

AGENCY

The agency is the piece of equipment or object closest associated with the loss and is the substance, object, or radiation most closely associated with the accident's consequence. The agency could be the principal objects, such as tools, machines, or equipment, involved in the accident and is usually the object inflicting injury or property damage (Figure 6.4).

The agency is therefore a key factor in the exchange of energy. It is the agent that is responsible for the energy transfer to the recipient, who consequently suffers a loss in the form of injury, occupational illness, or disease. Or, in the case of a property damage accident, damage to equipment, business interruption, or both.

TWO TYPES OF AGENCIES

There are two major classifications of agencies that are involved in the exposure or exchange of energy. They are *occupational hygiene agencies* and *general agencies*.

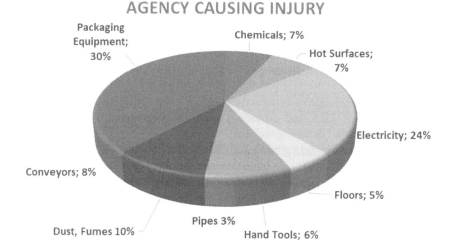

FIGURE 6.4 An analysis of agency types.

Occupational Hygiene Agencies
Occupational hygiene agencies include the following:

- Chemicals
- Dust
- Fumes/vapors
- Noise/vibration
- Heat/cold
- Fire/smoke
- Gas/fumes/vapors
- Radiation
- Fungus/mold
- Biological
- Ergonomics
- Illumination
- Ventilation
- Other

Occupational hygiene agencies are those items that are closest to and cause illness or occupational disease. An occupational disease or illness as a result of an accident is classified in the same way as an injury.

General Agencies
General agencies include the following:

- Walkways
- Ladders
- Sharp edges
- Power tools
- Hand tools
- Equipment
- Animals
- Boilers and pressure vessels
- Prime movers and pumps
- Radiation and radiating substances
- Chemicals
- Conveyors
- Electrical apparatus
- Elevators
- Machinery
- Highly flammable and hot substances
- Hoisting apparatus
- Power transmission equipment
- Working surfaces
- Other

AGENCY PART

The agency part is that part or area of an agency that inflicted the actual illness, injury, or damage. For example, a worker was ripping planks on a circular saw. To speed up production, he removed the machine guard, thus exposing the blade. During the cutting process, he was distracted, and the blade cut his finger. In this case, the agency is the circular saw, and the saw blade would be classified as the agency part.

AGENCY TRENDS

Trends can be used to establish which agency is responsible for the majority of losses as a result of an energy transfer. A trend analysis can be made by listing the agency that causes the injury, disease, or damage.

ANALYSIS OF BODY PART INJURED

Injury analysis will show which part of the body is being injured more frequently. This analysis helps determine which types of energy transfer to what areas of the body are more frequent. Although this is an injury analysis exercise, it does act as a measure for future accident prevention efforts to reduce accidental energy transfers, specifically to the body parts most often injured. This measurement may also indicate weaknesses in the personal protective equipment program.

OTHER SAFETY PERFORMANCE METRICS

There are other safety performance indicators, such as those used for road safety and airline safety.

ROAD SAFETY PERFORMANCE MEASUREMENTS

The following measures are commonly used to gauge road safety performance. Vehicle miles traveled (VMT) are quoted but can be substituted with vehicle kilometers traveled (VKT):

- Number of fatalities: The total number of persons suffering fatal injuries in a motor vehicle crash during a calendar year.
- Rate of fatalities: The ratio of the total number of fatalities to the number of VMT (in 100 million VMT) in a calendar year.
- Number of serious injuries: The total number of persons suffering at least one serious injury in a motor vehicle crash during a calendar year.
- Rate of serious injuries: The ratio of the total number of serious injuries to the number of VMT (in 100 million VMT) in a calendar year.

FATAL AND SEVERE INJURIES

To obtain the FSI (fatal and severe injuries) MVKT (vehicle crashes per million kilometers traveled) figures, calculate the sum of the total kilometers traveled over the crash period and the sum of FSI crashes. Kilometers can be substituted with miles. The following formula can be used:

$$\text{FSI MVKT} = \frac{\text{Sum (FSI crashes)}}{\text{Sum (Kilometers traveled during crash period)}} \times 100,000,000$$

AIRCRAFT ACCIDENTS

The fatal accident rate (FAR) used in the commercial scheduled airline industry is the number of onboard deaths per 100,000 flight hours. The formula is:

$$\text{FAR} = \frac{\text{Sum of onboard deaths}}{\text{Total flight hours}} \times 100,000$$

COSTS OF INJURIES AND ILLNESSES

All accidental injury and damage events cost money in the form of direct and indirect costs, which will be discussed in Chapter 11.

CONCLUSION

There are many statistics and tabulations that can be recorded, analyzed, and produced, based on injuries caused by accidents. All of these give a clear indication of what occurred in the past but tell little about the effort to reduce workplace hazards.

As part of the SMS processes, these loss statistics should be recorded and tabulated on a monthly and annual progressive basis. An ideal exposure period is expressed as a 12-month moving average. The figures should be presented in graphic format and tabled at management meeting as well as at board level.

Measuring a comprehensive health and safety management system by means of a correctly executed audit process will examine and test some 80 safety programs, processes, procedures, and standards, of which only a few relate to injury and non-injury statistic recording and analysis. The audit is a true measurement of proactive safety management performance. To measure safety management performance accurately, leading as well as lagging indicators must be monitored and measured.

SUMMARY

Fatality and injury rates are seemingly the only evidence of safety that is produced and published on a regular basis. These numbers are what legal bodies, management, and the public associate with safety. Lagging indicators are obtained after the event has occurred. An accident must occur and cause illness, injury, or property damage

Lagging Indicators: Injuries, Illness, and Diseases 77

before a statistic can be compiled. If there are no injuries as a result of an absence of accidents, the organization tallies an injury-free period as a *safety record*. This lagging indicator centers around the occurrence or nonoccurrence of an event, such as an accident.

The lost-time injury is perhaps the most widely used measure of safety performance. Most countries use the lost-time injury rate as a prime safety metric. This classification of injury distinguishes it from minor injuries in that the injury is so severe that the injured person loses time away from work. A lost-time injury (or disabling injury) results in temporary total disability. The definition of a lost time injury was modified to require the injured person to return to the *task he or she was doing at the time of the accident*. If the injured person was loading boxes onto a rack at the time of the accident, could they return to loading the same boxes on the same rack after the injury without losing a work shift?

Injury frequency rates have come to be recognized as an accepted indicator of safety management performance. Many are familiar with the term injury rate, and by using this injury rate, comparisons can be made with other industries of a similar type. There are other safety performance indicators, such as those used for road safety and airline safety.

7 Underreporting of Injuries, Illnesses, and Damage

MANIPULATION OF INJURY DEFINITIONS

Lagging indicators, which are dependent on the interpretation of disabling or recordable injuries, are unfortunately open to manipulation in an effort to maintain the safety record or avoid closer scrutiny by legal health and safety entities. Since most of the lagging indicators center around the interpretation of the severity of injury, the criteria are often manipulated. Terms such as "normal work" or "regular established job" are used to the organization's benefit, and injured workers are transferred to light duty assignments, which are then classified as a "regular established job."

Being "away from work" as a result of a workplace injury is also adverted by having the employee report to the worksite every day, irrespective of what work he or she would do there. Some organizations arranged special offices where injured employees would spend the work shift seated in front of computer screens. When questioned, the answer was that the employee was not away from work.

So much emphasis is based on the interpretation of lost time, disabling injuries, and recordable injuries that safety departments have been pressurized to keep the injuries low and have also resorted to interpretation manipulation.

CREATIVE BOOKKEEPING

In his paper, *Fixing the Workplace and not the Worker*, Bill Hoyle (2005) says,

> Workplace safety is measured by only one statistic, the OSHA recordable rate. Based on this statistical yardstick, continuous process industries continue to be among the safest industries in the country and are getting safer. Many plants have celebrated working millions of hours without a lost work-day accident. While all eyes are on the OSHA recordable rate, releases of hazardous materials, fires, mechanical breakdowns and near misses are not included in the safety statistics.

Bill Hoyle continues,

> Many workers know that the scoreboard injury numbers are often the result of creative book-keeping and of assigning injured workers special light duty work while they heal. The numbers may be a fantasy, but the jackets are real, so complaints are kept to a minimum.

p. 5

UNDERREPORTING OF INJURIES

The Health and Safety Executive (HSE) (UK) (2006) estimates that only half of the recordable injuries are reported,

> It is known that employers substantially under-report these non-fatal injuries: the level of overall employer reporting of RIDDOR defined nonfatal injuries to employees is estimated at around a half.

HSE Website

UNITED STATES GOVERNMENT ACCOUNTABILITY OFFICE

A report by the United States Government Accountability Office (GAO) (2009) concluded that:

> Workers, employers, and health practitioners all experience pressure to avoid reporting work-related injuries and illnesses. These stakeholders stated that workers' fear of disciplinary action and safety incentive programs that reward low injury rates are common disincentives for reporting. The report also notes widespread reports from occupational health practitioners who were pressured not to record an injury or illness.

> Some occupational health practitioners say that to avoid recording an injury, some employers will try to limit treatment for a serious injury to just first aid.

> In other cases, the practitioners said, employers might seek alternative diagnoses if the initial diagnosis would result in a recordable injury or illness. One manager took an injured worker to several medical providers until the manager found one who would certify that treatment required only first aid, thus making it an injury that did not have to be recorded, one practitioner told researchers, according to the report.

> Many employers fear that reporting numerous injuries will prompt a full-scale OSHA inspection. The accountability office said that 53 percent of health practitioners had reported experiencing pressure from company officials to play down injuries or illnesses, and that 47 percent had reported experiencing this pressure from workers.

According to the GAO report,

> Sixty seven percent of the 1,187 occupational health practitioners surveyed had reported observing worker fear of disciplinary action for reporting an injury or illness, and 46 percent said this fear had some impact on the accuracy of employers' injury and illness records. (GAO Website 2021).

pp. 1–35

HIDDEN TRAGEDY

During a hearing before the U.S. House of Representatives titled *Hidden Tragedy: Underreporting of Workplace Injuries and Illnesses* in Washington DC, on June 19, 2008, the report reads:

> This flawed system gives employers an incentive to underreport injuries. The fewer injuries and illnesses an employer reports, the less likely it will be inspected by OSHA

and the more likely it will pay lower premiums for workers' compensation. There is also mounting evidence that a number of employers are engaging in intimidation in order to keep workers from reporting their own injuries and illnesses.

A recent *Charlotte Observer* investigation on the hazardous working conditions in North Carolina's poultry industry revealed a shocking record of worker abuse and exploitation, often leading to crippling injuries and illnesses. The Observer also uncovered concerted efforts to discipline, intimidate and fire workers in retaliation for reporting serious on-the-job injuries. The Observer found that workers were forced to return to work immediately after having surgery so that the company would not have to file for workers' compensation.

Website (2022)

The International Labor Organization (ILO) (2000) reports that:

Under-reporting of occupational accidents and diseases is widespread, although the number of accidents and diseases that go unreported is difficult to quantify. Evidence has emerged to demonstrate that the scale of under-reporting is alarming. In 1990, the Health and Safety Executive of the *United Kingdom* sponsored a supplement to the 1990 Labor Force Survey containing questions on workplace injuries and ill health in order to establish the true level of workplace injury and of work-related ill health, and also to confirm the degree of under-reporting and the relative risk in the main industries. The findings showed that in the case of workplace injuries reportable to a safety authority, employers reported less than a third, and self-employed persons less than one in twenty.

ILO Website (2000)

THE SAFETY FEAR FACTOR

One of the biggest obstacles to safety efforts and the prevention of accidents is the fear factor that surrounds all aspects of safety at the workplace. Unless this fear factor is identified and the root causes of it eliminated by changing the safety philosophies of the organization, all efforts to honestly report injuries as a result of workplace accidents will fail.

The safety fear factor is also instilled in line management because their safety performance is measured only by injury rates. They want a good rating, so they underreport injuries. Employees are also judged harshly on their injury experience and undoubtedly hide injuries to prevent repercussions. International experience has shown that employees are under the impression (false or true) that they may lose their jobs as a result of being injured in an accident. This perception leads to injuries not being reported.

Some companies have a *three strikes and you're out* (unwritten) policy. This means that on the third accidental injury, the employee may be dismissed even though this is not legal. Some of the underlying reasons for the safety-fear factor are:

- Fear of disciplinary action from supervisors.
- Fear of being labeled incompetent, lazy, a complainer, or weak by coworkers and supervisors.

- Peer pressure because of poorly designed incentive programs.
- Preserving the safety record.
- Complicated and time-consuming paperwork.
- Lack of paid sick leave.
- It's just part of the job mentality.
- In extreme cases, fear of being laid off or let go entirely.

IMBEDDED CULTURE

Few safety professionals and fewer safety researchers will freely admit to the existence of a safety fear factor at workplaces in commerce, mines, and industry. Fear is imbedded in most organizations when it comes to safety, as safety is seen and gauged by the number of injuries that occur as a result of accidents. If you happen to be injured, you have spoiled the safety record, and rightly or wrongly, you are in the spotlight. There seems to be an automatic finger-pointing exercise after an accident, and we always seek out a guilty party. Employees are terrified of being fired if they are injured in an accident. Even though this seldom happens in reality, the paradigm remains imbedded in their minds.

INJURED WORKERS UNDER PRESSURE

All of these things can sometimes cause employers and managers to put a tremendous amount of pressure on an injured worker to cooperate in "hiding" their workplace injury. Employees are told: It's your fault, it will go on your record.

INTERNATIONAL CULTURE

There is obviously a culture of injury reporting suppression and injury manipulation in order to maintain a lower injury rate and achieve good safety performance. The fear factor and the blaming of employees for accidents are international occurrences. Most countries still gauge their industries' safety performance by the number of injuries experienced. Most legal entities that enforce safety also measure an organization by the injury rates. Inspections are triggered when a high number of injuries are reported. It is, therefore, no wonder then that the main focus is on injuries and inevitably the person who suffered the injury – commonly referred to as the person who caused the accident. It's easier to blame a worker for the accident than spending time and effort investigating the real causes.

NO BLOOD – NO FOUL

In numerous instances, if the accident (the event) does not result in an injury, it is largely ignored as it did not affect the safety record. It's a case of no blood – no foul! Yet, under slightly different circumstances, the event may have caused serious injury. What seems to matter most is the resultant injury, and this is where the safety attention and focus lie.

Near Miss Incidents

The same barriers that exist with the reporting of workplace injury accidents exist when it comes to reporting of near miss incidents. What increases the apathy toward reporting of near miss incidents is that nothing happened. There was no injury, no property damage, and little or no interruption to the work at hand.

Few organizations have formal, structured near miss reporting programs, and fewer educate and encourage employees to report near miss incidents. The fear factor exists with the reporting of any occurrence that could perceivably get the worker into any kind of trouble.

Reporting Property Damage Accidents

Since there is resistance to report injury-producing accidents, there is automatically resistance to report property damage accidents. If no one was injured, why report it? If I report it, I will be asked what I have done about it. These are some of the comments from employees when asked why they do not report these events. A structured, formal reporting system should form part of the organization's safety management system, and employees should feel they can report without repercussion or reprimand. These downgrading events are vital clues to management and indicate where control systems are breaking down and where more effort should be placed to reduce the risks.

Safe Space

There must be a climate of trust between employees and management. This includes declaring a truce and moving the focus away from injury blame-fixing and fault-finding to a safe space where injuries, damage accidents, and near miss incidents can be reported without fear or reprimand. Safe space is a space where employees' safety concerns can be freely expressed without fear of ridicule.

SAFETY BRIBERY

In an all-out effort to reduce the injury rate (not the risk), some employers came up with a safety bonus program. They offered employees a bribe in the form (often) of hard cash if the employees were *safe* and did not suffer any injury during the prescribed period. If managers want zero injuries, they will get zero injuries, especially if they are *paying* for zero injuries.

Sometimes this bonus is in the form of a production bonus, which also has the effect of subconsciously forcing the worker to work quicker to meet the production targets. Working faster creates a high-risk situation, once again driven by money.

Injury-Free Bonuses

Paying a safety bonus (injury-free bonus, to use the correct term) lulls management into thinking they are running a safe organization, meanwhile the injuries are being driven underground so that bonuses can be received. This culture of safety bribery will prove to be one of the biggest obstacles to changing the safety culture at the workplace.

According to Hoyle (2005),

> If you get to a safety target, for example two million hours, everyone gets prizes such as a sharp safety award jacket. Gainsharing and variable pay awards are also based on having low injury statistics. Prizes and awards are behavior modification techniques called positive reinforcement. Safety consultant Thomas Smith explains that "The research shows that positive reinforcement is as bad as negative reinforcement. It's just a different side of the same coin. The ultimate goal is to control people."
>
> p. 17

Hoyle continues the discussion this time referring to the peer-pressure created by bonuses,

> Many workers know that the scoreboard injury numbers are often the result of creative book-keeping and of assigning injured workers special light duty work while they heal. The numbers may be a fantasy, but the jackets are real, so complaints are kept to a minimum. Workers are also keenly aware of the powerful peer pressure to not report injuries. Reporting an injury means you will be blamed by your peers for taking money and prizes out of the pockets of your co-workers. Million-hour jackets, gain-sharing awards and safety scoreboards are a favorite part of management's behavior modification toolbox. Management recognizes that the problem behavior that they are most concerned with is to entice workers to think twice before reporting injuries.
>
> p. 17

THE ROOT OF EVIL

Money paid to be injury-free is the root of the problem. Employees hide injuries, lie about injuries, and do all in their power to cover up injuries, all for the money. Paying employees a safety bonus is tantamount to encouraging them to be dishonest about their injuries and to purposefully hide them so that they can collect the monthly, so-called, *safety* bonus. This does not make employees work safer. It makes them cleverer at hiding injuries and deceiving the system.

As McKinnon (2007) puts it:

> Paying people to be injury-free drives injuries underground. Disciplining people for being injured also drives injuries underground. Removing certain privileges and openly ridiculing injured employees drives injuries underground. With all these injuries going unreported, management is lulled into a false sense of security and thinks the plant is safe. Meanwhile, under the surface, you have the walking wounded running the factory.
>
> p. 74

SAFETY INCENTIVE SCHEMES

Safety incentive schemes based on reactive metrics are aimed at preventing the injury – not the event. All too often, safety managers find themselves unwillingly inheriting an old-school safety incentive program based on trailing indicators. These programs reward employees working for a period of time without reporting injuries. While initially sometimes achieving dramatic injury reductions, these programs quickly deteriorate into a "self-perpetuating nightmare," as one safety manager put it.

SAFETY AWARDS

Many of these schemes boast about their achievements. "With these incentives, workers get a sweatshirt for 250,000 workhours worked without a lost-time accident, a jacket for 500,000 workhours, and a $500 savings certificate and wall clock for 700,000 workhours."

Another organization gives gifts and individual recognition to associates who achieve an "accident-free year" (meant to read *injury-free* year). Another scheme even boasts of giving its employees a day's paid vacation for every six months worked without an accident. (They mean without *disabling injury*.)

INJURY-FREE PERIODS

An employee who had been injured in an accident was interviewed after he had tripped, fallen, and broken his ankle. After the accident, the employee continued to work for six hours on his broken ankle. Eventually, he was so overcome by the pain that he collapsed and had to receive medical attention. His ankle was so swollen that the attendants had to cut off his boot with a pair of tin snips. He was reluctant to report the injury and spoil the safety record. Is this loyalty, or an example of the fear of reporting an injury, letting the side down, and crashing the "safety record?" Injury-free periods need not necessarily indicate good safety performance but may indicate clever nonreporting or underreporting.

FALSE IMPRESSION

Impressive periods without a serious injury being recorded give management a false sense of security. They are lulled into thinking the workplace is safe because they are not experiencing accidental injuries. Since they are under the wrong impression, efforts and actions to constantly identify and reduce risks in the workplace are neglected. A situation of safety complacency is created by this seeming absence of work injuries.

BASED ON ACTIONS

Lagging safety targets are based on the measurement of unknown consequences. They are after the event, whereas safety targets should be based on actions before the event. Once the accident has been set in motion, the consequences cannot be accurately determined.

SAFETY GIMMICKS

Safety gimmicks are outdated, old-fashioned attempts at improving workplace safety and are sometimes an embarrassment to the profession. An example from *Changing Safety's Paradigms* (McKinnon 2007) follows:

> The Cheerleader Squad
>
> Another story is that of a company that called in the cheerleaders. Apparently, they were experiencing a high injury rate and once again the safety department was instructed to "do something about it."
>
> The lady relating the story was the safety coordinator at that plant, and she told us that they had recruited a team of cheerleaders. In full reggae, these cheerleaders then pounced on each department and posed for photographs with the operators, mechanics, and other workers in the field. The resultant photograph pictured smiling, happy cheerleaders and surly workers posing together.
>
> I cannot find the connection between the cheerleaders and safety, but apparently this was termed a "safety campaign." The visit by the cheerleaders was, I suppose, intended to cheer up the workers, or change their attitude or something. What I do know, however, is that it seemed to be a ridiculous thing to do. How could cheerleaders visiting the workplace possibly improve the safety? Wouldn't it have been far more effective for management to have visited the workplace? What was the purpose of taking photographs, posing with the cheerleaders? If this is safety promotion, then it has failed. What it has done, however, is make safety and safety promotion schemes appear more ridiculous in the eyes of the average worker.
>
> p. 172

SAFETY PUBLICITY BOARDS

Over the years, safety practitioners have managed to convince organizations to erect massive safety billboards, on which the safety record is proudly displayed. "This organization has worked X days without an accident" (Should be X days without an *injury*.) The safety billboard is an old-fashioned advertising gimmick that is still used to promote safety today. Other billboards read, *Date of last disabling injury* or *Department in the doghouse*. Here, the name of the department responsible for the last injury is displayed in the doghouse. On some safety billboards, a picture of a doghouse is displayed. Sometimes the injured person's name appears in the doghouse. This is an incentive for non-reporting; after all, who would want their name to appear in the doghouse?

INCORRECT TERMINOLOGY

It is also clear from the terminology that these schemes do not know the difference between an *injury* and an *accident*. They refer to *accident free* when they mean *injury-free*, which also leads to confusion.

EVIL RECORD

The safety record is actually an evil record as it encourages non-reporting and underreporting all for the sake of keeping the record intact. Rather than encouraging workers to work safer, it encourages workers to be injury-free. The focus of the record is no longer safety but having no injuries. The result of this is that low or zero injury rates are achieved by all means possible and not by good safety management.

ZERO HARM

The health and safety profession is constantly looking for new ways to improve safety in the workplace. Rather than focusing on the basics of hazard identification and risk reduction, organizations are swept up by campaigns and targets that sound good, are full of good intentions, but are unachievable and unrealistic.

A Goal

A goal is an achievable outcome that is typically broad and long-term. A company might use goals to inform the yearly strategies that each department will execute. *Zero Harm* is therefore an organization's goal. Whether *Zero Harm* is an achievable outcome is debatable.

An Objective

An objective defines the specific, measurable actions each manager and employee must take to achieve the overall goal of Zero Harm. The main difference between a *goal* and an *objective* is that goals provide direction, whereas objectives measure how you should follow that direction. A strategy defines how each manager, employee, or division will accomplish the objective.

Strategy Must Be Defined

In occupational health and safety, an *objective* refers to the specific *steps* a company will take to achieve a desired result. The result is the *goal*. *Zero Harm* is a goal that is stated, but it often lacks the objectives or specific steps that will be taken to achieve this goal. Seldom is the strategy defined, documented, and converted into actions.

The Zero Harm goal is to cause no harm to employees, contractors, or visitors at a place of work. The objective is how the organization plans to get to Zero Harm. The objectives needed to achieve Zero Harm are seldom defined; if they are not defined and if standards of accountability and measurement are not established for the objectives, they can't be managed.

The Health and Safety Management System (SMS)

The SMS has goals and objectives and has strategies for the achievement of these objectives. Each objective of the SMS should be measurable and achievable through the application of good management practices.

CAMPAIGNS

Buzzword campaigns like *Zero Injuries, Zero Harm* and *Zero Harm and Beyond*, sound good and profess an honest intent to prevent workplace injuries, but they are not realistic and are unattainable. Organizations cannot set targets that are unrealistic and unattainable. All agree that zero harm is the ultimate goal, but the goal is a consequence of an accident. More focus should be put on preventing the event (the accident) that caused the harm (consequence) as the consequence is often determined by fortuity. Many realize that zero harm is impossible to achieve. The concept *Zero Risk* is also impossible to achieve, as there is always residual risk in every walk of life. If zero risk is impossible to achieve, what about zero harm?

Many argue that a company adopting a zero-harm culture shows their dedication to safety. If they want to show real dedication to safety, why not then embark on a zero-risk campaign? After all it's the risk that causes the harm. Focusing on harm puts the cart before the horse.

Paper Cut Injury?

Zero harm is also a go-no-go measurement that does not give leeway for measuring performance or progress towards the objective. It's a matter of harm or no harm. One seminar attendee asked a question that silenced the room, "What about the receptionist who suffered a paper cut to her finger? Does that ruin the zero-harm record?"

Many say that focusing on zero harm is dreaming the impossible dream. A possible dream is implementing a fully functional health and safety management system aimed at identifying hazards and reducing risk to a tolerable level.

CONCLUSION

Measuring safety, or gauging whether a workplace is safe, based solely on injury tallies and injury and severity rates is unreliable, inaccurate, and should not be seen as a true representation of the state of safety. Injury and severity rates and their accompanying costs are lagging indicators of safety and are based on after the fact. They tell little of the safety effort of the organization and should therefore be used in conjunction with leading indicators of safety, as indicated by the efforts to reduce workplace risk via the interventions provided within a comprehensive health and safety management system (SMS).

SUMMARY

Lagging indicators, which are dependent on the interpretation of a disabling or recordable injury, are unfortunately open to manipulation in an effort to maintain the safety record or avoid closer scrutiny by legal health and safety entities. So much emphasis is based on the interpretation of lost time, disabling injuries, and recordable injuries that safety departments have been pressurized to keep the injuries low and have also resorted to interpretation manipulation.

The safety fear factor is also instilled in line management because their safety performance is measured only by injury rates. They want a good rating, so they underreport injuries. Employees are also judged harshly on their injury experience and undoubtedly hide injuries to prevent repercussions. International experience has shown that employees are under the impression (false or true) that they may lose their jobs as a result of being injured in an accident.

There is obviously a culture of injury reporting suppression and injury manipulation in order to maintain a lower injury rate and achieve good safety performance. The fear factor and the blaming of employees for accidents are international occurrences. Since there is resistance to report injury-producing accidents, there is automatically resistance to report property damage accidents. If no one was injured, why report it?

In an all-out effort to reduce the injury rate (not the risk), some employers came up with a safety bonus program. They offered employees a bribe in the form (often) of hard cash if the employees were *safe* and did not suffer any injury during the prescribed period. If managers want zero injuries, they will get zero injuries, especially if they are *paying* for zero injuries.

Paying employees a safety bonus is tantamount to encouraging them to be dishonest about their injuries and to purposefully hide them so that they can collect the monthly, so-called, *safety* bonus.

8 Awards Based on Lagging Indicators

Traditionally, safety performance was gauged by the fatality and injury rates. Seemingly, this was the logical way to measure the safety performance of organizations and to draw comparisons with similar industries and national averages. National awards and recognition are given to companies that have a perfect injury-free record or that display improvements in their injury rate in comparison with previous years.

THE NATIONAL OCCUPATIONAL SAFETY ASSOCIATION (NOSA)

The National Occupational Safety Association (Section 21) (NOSA) of South Africa was established in 1951 to promote the prevention of occupational accidents and diseases on a national scale and to act as an advisory body on all aspects of workplace safety.

COLLECTION OF INJURY STATISTICS

Through its network of 12 regional offices, monthly injury statistics were collected from industries and mines falling within each regional area. These were mailed in to NOSA regional offices on a special form, and the figures were recorded. A new form, which contained the current injury frequency rate derived from the submitted statistics, was mailed back to the employer, informing them of the progressive injury rate and also providing a new form for the submission of the next month's statistics.

NOSA AWARD PLAN

At the end of each year, the regional offices totaled the injury figures for each submitting workplace and determined if they were eligible for an award based on their injury statistics.

A Certificate of Commendation was awarded to a company that had worked 50,000 injury-free workhours or more during the calendar year. A Merit Perfect certificate was awarded to companies that had achieved one million or more injury-free workhours, and an Honor Perfect was awarded for three million or more injury-free workhours.

Based on an improvement formula, three years' injury statistics were compared with the current injury rate, and if the improvement was sufficient, a Merit Betterment award was issued. An Honor Betterment award was given if the company exceeded the minimum improvement requirement.

The awards were then presented to the award-winning companies at a national safety awards ceremony. The collected statistics were also used to calculate the national injury frequency rate of member companies on a national basis.

NATIONAL SAFETY COUNCIL (US) AWARDS

The National Safety Council (NSC) (US) has a number of awards for injury-free periods and improvements in injury rates. These awards are based on self-reporting by the organizations.

SUPERIOR SAFETY PERFORMANCE AWARD

This award recognizes organizations that have achieved ten or more consecutive years without an occupational injury or illness resulting in days away from work. To qualify for this award, the organization must achieve:

- Zero work-related illnesses involving time away from work within a consecutive 10-year period.
- Zero injuries involving time away from work in a consecutive 10-year period.
- Zero fatalities in a consecutive 10-year period.

SIGNIFICANT IMPROVEMENT AWARD

The Significant Improvement Award recognizes organizations that have a 20% or more year-over-year reduction in employee injuries and illness. The following are the qualifying criteria:

- A 20% reduction in injuries and illnesses over the previous calendar year.
- The maintenance or reduction in the number of fatalities over the previous calendar year.
- The maximum allowable injuries and illnesses for eligibility are at the discretion of the NSC.

PERFECT RECORD AWARD

The Perfect Record Award recognizes individuals, companies, units, or facilities that have completed a period of at least 12 consecutive months without incurring an occupational injury or illness that resulted in days away from work or death.

THE MILLION WORKHOURS AWARD

This award recognizes organizations that have completed a period of at least one million consecutive workhours without an occupational injury or illness resulting in days away from work. The criteria include:

- Zero work-related illnesses involving time away from work for at least one million consecutive work hours.

- Zero injuries involving time away from work for at least one million consecutive work hours.
- Zero fatalities for at least one million consecutive work hours.

MINING AWARDS

South African mines have annual awards based on fatality-free achievements. The Millionaire Award is an award to recognize the milestone of achieving one million fatality-free shifts on any mine. The Thousand Fatality-free Production Shifts award is an award to recognize the milestone of achieving 1,000 fatality-free production shifts on any mine. The Safety Achievement Flag is awarded to the mines that have the highest percentage improvement in their allocated risk (days lost when comparing two consecutive three-year periods).

THE NOSA 5-STAR SAFETY AND HEALTH MANAGEMENT SYSTEM

The NOSA 5-Star Safety and Health Management System was developed by the NOSA in South Africa in about 1968. The original score sheet was developed to generate interest in safety among a group of timber mills and sites. A checklist of 20–30 high-risk elements found at lumber sites was compiled, and a score (or weighting) was allocated to each element so that by scoring on the checklist, they could compare leading safety efforts on their various sites. The original idea was to have a competition between sites but not use the injury rate as the only measurement. The score allocated after the inspection and ranking would indicate their safer sites.

This system was so successful that NOSA started conducting safety surveys based on this system at mines, industries, and commerce. After 50,000 safety surveys and inspections, more elements were added, and the NOSA safety management system (SMS) was then promoted among industries and mines throughout South Africa. Around 1970, the system consisted of 40 elements grouped under 5 main sections with a total score of 2000 points.

Star Grading Awards

Once organizations had implemented the safety system, NOSA awarded companies a rating, or grading, of their safety systems, in the form of an A, B or C grade based on the audit score. This proved very popular as organizations could now be recognized for their safety efforts based on management work being done to prevent loss, and not just on injury rates. The system used both leading and lagging indicators to award a star rating.

Safety Effort and Experience

With the advent of the hotel star grading system around 1975, NOSA linked the score to a star grading based on the SEE formula, *Safety Effort and Experience (SEE)*. A score was allocated for the safety effort (safety system effectiveness), and in conjunction with the injury rate (experience), a star grading from 1 to 5 was awarded.

STAR GRADING	EFFORT	DIIR	DIFR
5-Star	91%	1%	5
4-Star	74%	2%	10
3-Star	60%	3%	15
2-Star	50%	4%	20
1-Star	40%	5%	25

FIGURE 8.1 The table shows the injury rate and effort score required for a star grading.

Challenging organizations to achieve a 5-Star grading for safety effort proved to be highly successful, and eventually the NOSA 5-Star Safety System was exported to and used in no fewer than 10 countries. This was one of the first safety systems to measure proactive, upstream safety efforts. It was extensively used to grade health and SMS efforts and recognize them on a national and international scale by awarding a NOSA star grading (Figure 8.1).

The final effort score was taken in relation to the injury rate, which determined the star grading awarded. The effort score was determined after an audit of all the elements of the SMS. A star grading was based on the injury experience (lagging indicator) and safety efforts (leading indicator) of the preceding 12 months (Figure 8.1) and was valid for a year. The audit examined all the activities and processes that took place during the preceding 12 months as well.

OCCUPATIONAL SAFETY AND HEALTH ADMINISTRATION (OSHA) VOLUNTARY PROTECTION PROGRAM (VPP)

The OSHA VPP (voluntary protection program) also uses the safety effort and experience model to recognize organizations. The VPP recognizes employers and workers in the private industry and federal agencies that have implemented effective health and SMS and maintain injury and illness rates below national averages for their respective industries. In VPP, management, labor, and OSHA work cooperatively and proactively to prevent fatalities, injuries, and illnesses through a system focused on hazard prevention and control, worksite analysis, training, and management commitment and worker involvement.

THE VPP STAR PROGRAM

The Star Program is designed for exemplary worksites with comprehensive, successful safety and health management systems. Companies in the Star Program have achieved injury and illness rates at or below the national average of their respective industries. These sites are self-sufficient in their ability to control workplace hazards. Star participants are reevaluated every three to five years, although incident rates are reviewed annually.

The Star Demonstration program is designed for worksites with Star quality safety and health protection to test alternatives to current Star eligibility and performance requirements. Promising and successful projects are considered for changes to Star requirements. Star Demonstration program participants are evaluated every 12–18 months.

To obtain Star status, OSHA requires worksites to have successful ongoing health and SMSs, cooperation between labor and management, and work-related injury and illness rates at or below the average rate for their respective industries.

SUMMARY

Traditionally, safety performance was gauged by the fatality and injury rates. Seemingly, this was the logical way to measure the safety performance of organizations and to draw comparisons with similar industries and national averages. National awards and recognition are given to companies that have a perfect injury-free record or that display improvements in their injury rate in comparison with previous years.

The NSC has a number of awards for injury-free periods and improvements in injury rates based on lagging indicators. These awards are based on self-reporting by the organizations.

South African mines have annual awards based on fatality-free achievements. The millionaire award is an award to recognize the milestone of achieving one million fatality-free shifts on any mine.

The NOSA 5-Star Safety and Health Management System was developed by the NOSA in South Africa in about 1968. The original score sheet was developed to generate interest in safety among a group of timber mills and sites. A checklist of 20–30 high-risk elements found at lumber sites was compiled, and a score (or weighting) was allocated to each element so that by scoring on the checklist, they could compare leading safety efforts on their various sites.

With the advent of the hotel star grading system around 1970, NOSA linked the score to a star grading based on the SEE formula, *Safety Effort and Experience (SEE)*. A score was allocated for the safety effort (safety system effectiveness), and in conjunction with the injury rate (experience), a star grading from 1 to 5 was awarded.

The OSHA VPP also uses the safety and experience model to recognize organizations. The VPP recognizes employers and workers in the private industry and federal agencies that have implemented effective health and SMSs and maintained injury and illness rates below national averages for their respective industries.

9 Lagging Indicators of Safety Performance
Damage, Fires, and Interruption

ACCIDENT OUTCOMES

For years, employers focused on workplace injuries as an indicator that something was wrong with the system. All efforts were directed toward injuries, and all efforts to prevent them were focused on the injury and the event that caused it. Little, if any, attention was paid to what was happening below the waterline in the form of undesired events that were damaging equipment, tools, materials, and products. Often these losses went unreported and were never investigated, yet the difference between the same event causing damage or personal injury was a matter of timing or position, as OSHA puts it, or luck. as McKinnon (2000) puts it:

> The failure to identify hazards, assess the risks and institute the necessary controls causes accident root causes to exist. They in turn create high-risk behavior and high-risk conditions which, as Luck Factor 1 would have it, result either in a near miss incident (no loss) or an exposure, impact or contact with a source of energy (loss).
>
> The degree of exposure, impact or contact is what causes the loss. The outcome of an exposure, impact or exchange of energy is determined by fortuity or Luck Factor 2. The outcome may be injury to employees, damage to property, business interruption or a combination.
>
> p. 123

Once the exposure, impact, or exchange of energy takes place, the outcome cannot be accurately predicted. Once there is an exposure, impact, or exchange of energy, we have no control over the results.

The result of this exposure, impact, or exchange of energy, if above the threshold limit of the body or the substance, could be:

- Disease or illness.
- Personal injury.
- Property damage.
- Business interruption.
- Or a combination of all of the above. (p. 124)

UNINTENDED EVENTS

Accidental property damage and business interruption events do not include acts of sabotage or normal wear and tear. They are unintended and undesired events that result in damage to property and equipment or cause disruption due to an unplanned, uncontrolled exposure, impact, or energy transfer.

BRIDGING THE GAP

Bridging the gap between traditional injury prevention programs and health and safety management systems (SMS) meant the recognition, investigation, and reduction of accidents that resulted in property and equipment damage. In shifting the focus from the tip of the iceberg, i.e., the injuries, Bird, and Germain (1992) explain the concept:

> Third, if the event results in property damage or process loss alone, and no injury, it is still an accident. Often, of course, accidents result in harm to people, property, and process. However, there are many more property-damage accidents than injury accidents. Not only is property damage expensive, but also damaged tools, machinery and equipment often lead to further accidents.
>
> p. 18

ACCIDENTAL PROPERTY DAMAGE

All organizations that experience injury-resulting accidents are also experiencing accidents that damage equipment and property, as well as numerous near miss incidents. Focusing on the injury-producing accidents tends to take the focus away from the bigger picture of what is occurring in reality.

ACCIDENT RATIOS

In the 1930s, H.W. Heinrich proposed that there was a ratio between serious injuries, minor injuries, and narrow escapes (near misses) as he termed them. Although not all of Heinrich's theories are accepted as being true, more recent studies by reputable organizations and researchers have confirmed that there is a definite ratio of minor injuries to serious injuries and near miss incidents to serious injuries.

In their 1966 book *Damage Control: A New Horizon in Accident Prevention and Cost Improvement*, Frank E. Bird and George L. Germain make a strong case for investigating accidents, not just those cases that produce injuries. They say that the study of accidents instead of injuries does not downgrade the importance of preventing human injury. Rather, it recognizes that many "no-injury accidents" might have resulted in personal injury, property damage, or both.

In *Practical Loss Control Leadership* (1992), Bird and Germain clarify the reason why property accidents were never considered as serious as injury-producing accidents, the reason being that the term *accident* only refers to *injury-resulting* accidents.

> As we consider the ratio, we observe that 30 property damage accidents were reported for each serious or disabling injury. Property damage accidents cost billions of dollars

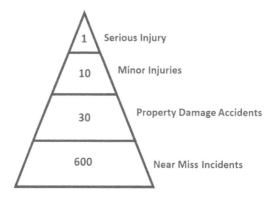

FIGURE 9.1 The Frank E. Bird and George Germain 1969 accident ratio.

annually and yet they are frequently misnamed and referred to as near accidents. Ironically this line of thinking recognises the fact that each property damaged situation could probably have resulted in personal injury. This term is a holdover from earlier training and misconceptions that led supervisors to relate the term *accidents* only to *injury*.

p. 21

ACCIDENT RATIO STUDY

Their 1969 accident ratio study, based on 1.75 million accidents, revealed that for every serious injury, there were ten minor injuries, 30 property damage accidents, and 600 incidents where there was no visible injury or damage (Figure 9.1).

According to DNV (2023), *A Tribute to Frank Bird:*

> At Lukens Steel, he headed a 7-year study (1959-65) of 90,000 incidents. It revealed a ratio of 1:100:500. That is, for every 1 disabling injury there were 100 minor injuries and 500 property damage accidents. Property damage accidents were prevalent and very costly, with a huge potential for harm to people. Interest in this "all-accident control" concept spread around the world thanks to presentations by Frank and others at national and regional safety and operating management conferences.

DNV Website (2023)

OTHER RATIOS RESEARCHED

This was followed by U.K. studies in 1972 by Fletcher, who studied 50 plants owned by one multinational company that operated in 12 countries.

In 1974–1975, the Tye–Pearson theory (Figure 9.2) was conducted on behalf of the British Safety Council and was based on a study of almost one million accidents in Britain. The ratio showed that for each serious injury experienced, three minor injuries occurred, 50 first aid injuries took place, 80 accidents caused damage, and there were in excess of 400 near miss incidents. The study was concluded by stating

Lagging Indicators: Damage, Fires, and Interruption

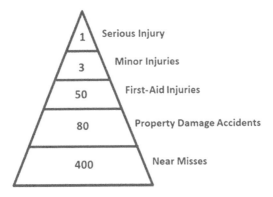

FIGURE 9.2 The Tye–Pearson accident ratio. (From the British Safety Council. 1974/1975. *Tye–Petersen Theory*. With permission.)

that there are a great many more near miss incidents than injury- or damage-producing ones, but little is generally known about these.

The Tye–Pearson accident ratio indicated that the sample workplaces studied experience more than 80 accidents which did not cause injury or ill-health to employees but which damaged equipment or property.

The accident ratio (Figure 9.3) was derived from a study by the Health and Safety Executive (HSE) of Great Britain in 1993. The HSE ratio shows that for every serious or disabling injury, 11 minor injuries were experienced, and 441 property damage accidents occurred.

CRITICISM OF ACCIDENT RATIOS

Many criticize well-researched accident ratios from accredited sources. The focus of the criticism is usually on the numbers quoted in the ratios. If an organization is having accidents, then they are more than likely having accidents that damage property or equipment or which interrupt production. Then they are also experiencing near miss incidents and therefore have a ratio between injury, damage, and near miss incident events. The numbers can be criticized, but the fact remains that there is a

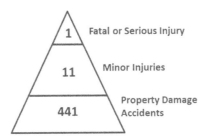

FIGURE 9.3 The Health and Safety Executive (HSE) accident ratio.

proportion of accidents that do not result in injury, but cause some form of damage or other form of loss.

Four principles based on the accident ratio are as follows:

- There are consistently a greater number of less serious injuries than more serious injuries.
- Accidental events occur that damage property and have the potential to cause injury.
- Many near miss incidents could have become events with more serious consequences.
- All the undesired events (not just those causing injuries) represent failures in control and are potential opportunities to prevent accidents.

INVESTIGATION AND RECORDING

Simply because the accident did not cause injury, it is not advisable to let the event go by without investigation. In risk management, a keyword is *potential*. Most accidents that damage property or equipment or which disrupt the business have sufficient energy transfer *potential* to have injured workers under slightly different circumstances.

RISK MATRIX

A simple risk matrix (Figure 9.4) can be used to determine the potential severity and the probability of recurrence of a property damage accident. Although there was no injury as a result of the event, the risk matrix will indicate the injury potential that the event had, as well as the probability of a similar event occurring.

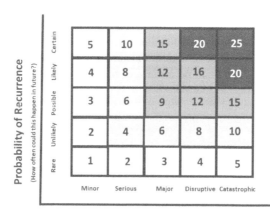

FIGURE 9.4 A risk matrix.

Any property damage accident that causes more than $1,000 worth of damage or disruption should be investigated with the same rigor as an injury accident.

CALCULATING THE ACCIDENT RATIO

A statistic that is perhaps the most significant when measuring safety management performance is the organization's own accident ratio. If the statistics are insufficient to compile a ratio, then the organization is not receiving all reports of:

- Serious or recordable injury cases.
- Minor injuries.
- Property damage accidents.
- Near miss incidents.
- Fires.
- Other disruptive occurrences.

These reporting systems must be effective for the organization to see the complete picture of losses being experienced. Nonreporting or under-reporting could be because of certain barriers.

BARRIERS TO REPORTING OF DOWNGRADING EVENTS

Because of the safety fear factor, many downgrading events may not be reported. These could be because:

- Employees don't know they are supposed to report property damage accidents; after all, no one was injured.
- Employees do not know how to go about reporting property damage accidents. The training was insufficient, or the reporting methods are not clear.
- They are afraid of being reprimanded or disciplined for actions that led to the property damage accident.
- Employees feel pressure from co-workers to keep quiet so that nobody gets into trouble and nobody loses the safety bonus or spoils the safety record.
- They are under pressure to maintain a clean property damage accident record.
- They are new to the crew and want to make a good impression. Why make waves and stand out in the crowd for nothing?
- The workplace safety culture says, "Nothing happened, no one was hurt so don't make a big deal out of it."
- Co-workers view the property damage accident with humor instead of seeing the hazard. If everyone is laughing, how serious could it be?
- Last time they tried to talk to the supervisor about a property damage accident, they were belittled or disregarded.

- It's just too much trouble to fill out those forms, and they have no time for paperwork.
- Property damage accident reporting is not encouraged by the organization.

AMNESTY

If management wants the reporting system to work and contribute to the reduction of risk and consequent losses to the organization, a decision to declare amnesty must be made. The reporting mechanism should allow for anonymity in reporting but also allow for the reporter to volunteer their name if they feel comfortable doing so, such as in the case of reporting safe work or deeds. Blame-fixing and punitive actions based on reports where the employee was obviously at fault must be avoided. This may be a difficult paradigm for most managers to change, as most link safety violations with punitive actions.

The employee grapevine works well, and the first time there is punishment leveled at an employee who reports a high-risk act by another employee or themselves, the reporting will dry up. Management must take a bold decision in this regard and accept that discipline in safety has never worked and never will.

As the safety icon, Frank E. Bird Jr., said:

> Punishment is the last resort, but it must be done in a way that communicates your genuine concern.
>
> p. 52

Frank Salas, Joint Union Management Safety Coordinator, once told me, "Discipline in safety hasn't worked for 20 years, why will it work now?"

VITAL STATISTICS

These statistics are vital for the calculation of the organization's accident ratio and for accurate measurement of safety performance and the costing of accidents.

CALCULATING AN ACCIDENT RATIO

To calculate the accident ratio for a period of 6 months, all the serious injuries must be totaled, as well as the minor injuries, property damage events, and near miss incidents. The serious injury figure is then converted to one by dividing by itself. The minor injury total is also divided by the number of serious injuries, as well as the property damage accidents and the reported near miss incidents. This would derive a ratio of one serious injury in relation to minor injuries, property damage accidents, and near miss incidents.

The following examples are used to derive an accident ratio:

- 15 serious injuries were experienced in the period.

$$\frac{\text{All serious injuries (15)}}{\text{All serious injuries (15)}} = 1$$

Lagging Indicators: Damage, Fires, and Interruption

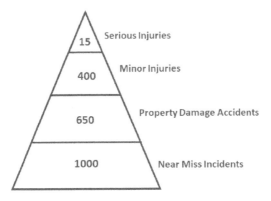

FIGURE 9.5 The organization's loss figures.

- 400 minor injuries were recorded in the same period.

$$\frac{\text{Minor injuries }(400)}{\text{All serious injuries }(15)} = 26$$

- 650 property damages (including fires) were reported.

$$\frac{\text{Property damage accidents }(650)}{\text{All serious injuries }(15)} = 43$$

- 1,000 near miss events were recorded during the period.

$$\frac{\text{Near-miss incidents }(1,000)}{\text{All serious injuries }(15)} = 66$$

Figure 9.5 shows the actual totals, and Figure 9.6 is the derived accident ratio for the organization. The organization's accident ratio for the period is: 1:26:43:66.

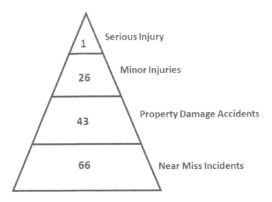

FIGURE 9.6 The organization's derived accident ratio.

The analysis of the accident ratio produced shows a high number of minor injuries per serious injury. It shows the correlation between property damage accidents and serious injuries as well as a low report of near miss incidents, although reporting is occurring. The goal of the accident ratio is to show that the number of near miss incidents should be reduced by effective safety management controls, which in turn will reduce the number of property damage accidents, minor injuries, and serious injuries. The base of the triangle must be tackled first before injuries and damage events can be reduced.

COSTING

Some of the major types of property damage could include:

- Buildings.
- Materials.
- Fixed equipment.
- Materials handling equipment.
- Motor vehicles.
- Tools.

A simple classification for the level of property damage for investigation purposes is:

1. Minor (less than $100).
2. Serious ($100–$1,000).
3. Major ($1,000–$10,000).
4. Catastrophic (over $10,000).

As stated, property-damage accidents are caused in exactly the same sequence as injury-producing accidents. As they also have the potential to cause injury, they should receive the same attention and investigation as injury-producing events. To determine the total cost of risk within an organization, the costing of all losses as a result of undesired events should be done monthly and on a progressive basis.

MEASURING PROPERTY DAMAGE ACCIDENTS

As with injury rates, the number and frequency of property damage events can be tracked and monitored. Some metrics for this are as follows:

- The accident ratio.
- Total number of property damage events per month.
- Total number of property damage events annually progressive.
- Vehicle damage accidents per month and per year.
- Vehicle damage accidents per 100,000 miles or kilometers driven.
- Cost of vehicle accidents.
- Cost of property and equipment damage.
- Environmental losses.

FIRE DAMAGE ACCIDENTS

Accidental fires exclude fires as a result of external threats such as terrorism or fires set by disgruntled employees. In the US alone, local fire departments responded to 1,338,500 fires in 2020. These fires caused 3,500 civilian deaths, 15,200 civilian injuries, and US$21.9 billion in property damage. The same report quotes the cost of all injuries as US$1,158.4 billion. Work-related injuries cost US$163 billion during the same year, rising to $167 billion in 2021.

Even in medium-sized organizations, property damage as a result of accidents should be tracked and charged to establish the full cost of these accidents.

ENVIRONMENTAL HARM

Pollution of land, air, and water due to accidental discharges or emissions is regarded as an accident. Although no injury or work-related illness was experienced, the undesired event caused harm to the environment. All environmental harm due to accidents costs money and should be prevented by proactive actions and processes within the SMS. Measuring safety performance by injury counts alone does not give a full picture of management failures, which should include all property damage events, including those that affect the environment.

Exxon Valdez Oil Spill

The Exxon Valdez oil spill was a manmade disaster that occurred when *Exxon Valdez*, an oil tanker owned by the Exxon Shipping Company, spilled 11 million gallons of crude oil into Alaska's Prince William Sound on March 24, 1989. The ship struck Bligh Reef, a well-known navigation hazard in Alaska's Prince William Sound, and the impact of the collision tore open the ship's hull, causing some 11 million gallons of crude oil to spill into the water. It was the worst oil spill in U.S. history until the Deepwater Horizon oil spill in 2010. The Exxon Valdez oil slick covered 1,300 miles of coastline and killed hundreds of thousands of seabirds, otters, seals, and whales. Nearly 30 years later, pockets of crude oil remain in some locations.

OTHER MAJOR ACCIDENTS

The Chernobyl nuclear accident cost around US$200 billion. The loss of the space shuttle Columbia cost US$18 million on direct costs of the accident investigation board and US$112 million in supporting the board's investigation. The Prestige oil spill cost US$12 billion. The oil spill resulting from the Deepwater Horizon cost US$61.6 billion. The Bhopal accident cost US$470 million in settlements alone.

TOTALLY HIDDEN COSTS

Each accident results in some form of loss, and all losses cost money. Time may be lost, forms need to be filled out, and the business is interrupted to a degree. Many of the costs of an accident are hidden and therefore go unnoticed. Direct costs or insured costs are normally the only costs associated with an accident and are the lesser of the two amounts. There are also a number of totally hidden costs that are not immediately apparent.

SUMMARY

For years employers focused on workplace injuries as an indicator that something was wrong with the system. Little, if any, attention was paid to what was happening below the waterline in the form of accidents and undesired events that were damaging equipment, tools, materials, and products.

Bridging the gap between traditional injury prevention programs and health and SMS meant the recognition, investigation, and reduction of accidents that resulted in property and equipment damage.

In their 1966 book *Damage Control: A New Horizon in Accident Prevention and Cost Improvement*, Frank E. Bird and George L. Germain make a strong case for investigating accidents, not just those cases that produce injuries. They say that the study of accidents instead of injuries does not downgrade the importance of preventing human injury. Rather, it recognizes that many "no-injury accidents" might have resulted in personal injury, property damage, or both.

If an organization is having accidents, then they are more than likely having accidents that damage property or equipment or which interrupt production.

Property-damage accidents are caused in exactly the same sequence as injury-producing accidents. As they also have the potential to cause injury, they should receive the same attention and investigation as injury-producing events. To determine the total cost of risk within an organization, the costing of all losses as a result of undesired events should be done monthly and on a progressive basis.

In the US alone, local fire departments responded to 1,338,500 fires in 2020. These fires caused 3,500 civilian deaths, 15,200 civilian injuries, and US$21.9 billion in property damage. Pollution of land, air, and water due to accidental discharges or emissions is regarded as accidents. Although no injury or work-related illness was experienced, the undesired event caused harm to the environment.

10 Cause of Injury or Damage
Transfer of Energy

IMPACT, EXPOSURE, OR ENERGY TRANSFER TYPES

In every accident, there is some form of energy transfer that causes injury, property damage, or other forms of loss.

Once a high-risk behavior or a high-risk condition is imminent, there could be an exposure, impact, or exchange of energy. This exchange is in the form of contact with a substance or source of energy greater than the threshold limit of the body or article. The high-risk behavior or high-risk condition may result in a near miss incident, where, although there was a flow of energy and potential, there was no exposure, impact, or contact with a source of energy.

The exposure, impact or contact, or exchange of energy is the portion of the accident sequence that is most closely associated with the loss. The exposure, impact, or exchange of energy is what injures, damages, pollutes, or interrupts the business process.

ENERGY RELEASE THEORY

The energy release theory (Dr. Lestie Ball 1970) states that all accidents are caused by hazards, and all hazards involve energy, either due to involvement with destructive energy sources or due to a lack of critical energy needs. This model is most useful to identify hazards and understand system safety.

It was also proposed that injury to a living organism can only be caused by some energy exchange. Hence, it was suggested that the energy exchange should be considered as the injury or illness agent. The energy exchange resulting in an injury could be mechanical, chemical, thermal, electrical, etc. This concept is useful in understanding the way an injury is caused and in examining the solutions.

According to *SafetyRisk.net*, William Hadden's (1970) *energy release theory* portrays accidents in terms of energy transference. This transfer of energy, in large amounts and/or at rapid rates, can adversely affect living and nonliving objects, causing injury and damage. Thus, an accident is caused by energy out of control. The theory states that various techniques can be employed to reduce accidents, including preventing the buildup of energy, reducing the initial amount of energy, preventing the release of energy, carefully controlling the release of energy, and separating the energy being released from the living or nonliving object.

Haddon suggested ten strategies to prevent or reduce losses:

1. To prevent the initial marshalling of the form of energy.
2. To reduce the amount of energy marshalled.
3. To prevent the release of energy.
4. To modify the rate of spatial distribution of the release of energy from its source.
5. To separate in space or time the energy being released from the susceptible structure.
6. To separate the energy being released from the susceptible structure by the interposition of a material barrier.
7. To modify the contact surface, subsurface, or basic structure which can be impacted.
8. To strengthen the structure, which might be damaged by the energy transfer.
9. To move rapidly in the detection and evaluation of damage and to counter its continuation and extension.
10. All those measures that fall between the emergency period following the damaging energy exchange and the final stabilization of the process Haddon (2023).

ENERGY EXCHANGE – NOT ACCIDENT TYPES

Exposure, impact, or exchanges of energy are also (incorrectly) referred to as *accident types*. A more apt description would be *energy transfers*. Examples are:

- Struck against (running or bumping into).
- Struck by (hit by a moving object).
- Fall to a lower level (either the body falls or the object falls and hits the body).
- Fall on the same level (slips and fall, tip over).
- Caught in (pinch and nip points).
- Caught on (snagged, hung).
- Caught between (crushed or amputated).
- Contact with (electricity, heat, cold, radiation, caustics, toxics, noise, etc.).
- Over stress/over exertion/overload, etc.

The terms, *types of exposure, impact, contact,* or *energy exchange* are far more accurate descriptions than *accident types*, which was the term used in the past.

Energy Transfer Analysis

As part of the measurement of safety performance, a monthly and annual progressive analysis of the energy transfer types experienced should be produced. This will indicate trends in which energy transfer types are responsible for the losses. This is one form of lagging indicator measurement (Figure 10.1).

Cause of Injury or Damage

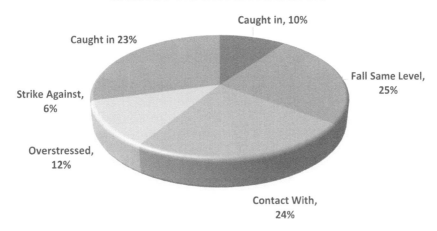

FIGURE 10.1 An example of an energy transfer analysis.

AGENCY

The agency is the piece of equipment, substance, object, or radiation most closely associated with the loss. It is the *principal* object, such as tools, machines, or equipment involved in the exposure and is usually the object inflicting injury or property damage.

Examples of agencies include:

- Animals.
- Boilers and pressure vessels.
- Chemicals.
- Conveyors.
- Dusts.
- Electrical apparatus.
- Elevators.
- Hand tools.
- Highly flammable and hot substances.
- Hoisting apparatus.
- Machines.
- Mechanical power transmission equipment.
- Prime movers and pumps.
- Radiation and radiating substances.
- Working surfaces and miscellaneous.

The agency is therefore a key factor in the exposure, impact, or exchange of energy. It is the agent that is responsible for the energy transfer to the recipient, who consequently suffers a loss in the form of injury or illness. The loss could also be damage or interruption.

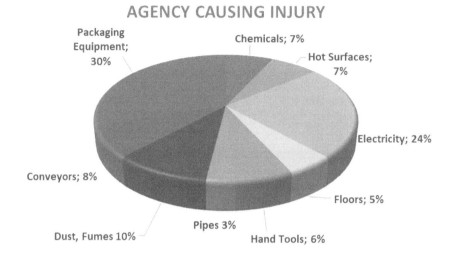

FIGURE 10.2 An example of an agency analysis.

AGENCY PART

The agency part is that part or area of an agency that inflicted the actual injury or damage. For example, a worker was ripping planks on a 12 inch (30 cm) circular saw. To speed up production, he removed the machine guard, thus exposing the blade. During the cutting process, he was distracted, and the blade cut his finger. In this case, the agency is the circular saw, and the blade would be classified as the agency part.

Agency Trends

A trend analysis can be made by listing the agency that causes the injury, disease, or damage. Trends can be used to establish which agency is responsible for the majority of losses as a result of the energy transfer (Figure 10.2).

CONTACT CONTROL

Management has three opportunities to exercise the safety management function of control within the accident sequence. The first opportunity is pre-contact control, which is setting up controls within the health and safety management system (SMS), identified by risk assessment, to manage the safety activities on a day-to-day basis. This reduces the root causes of accidents and, in turn, leads to a reduction in high-risk behaviors and high-risk conditions. This pre-contact effort will eliminate the accidental exposure, impact, or exchange of energy that causes the loss.

The second opportunity for control is during the contact (energy exchange) stage. Control during the contact stage of the accident sequence can only be directed at minimizing or averting the amount and type of energy exchange. Contact control does not stop the sequence of events that leads up to the exchange of energy. It only deflects or transfers the amount of energy in another direction so that it does not cause harm to the body or structure.

As Bird and Germain (1992) describe contact control:

> Control measures that alter or absorb the energy can be taken to minimize the harm or damage at the time and point of contact. Personal protective equipment and protective barriers are common examples. A hard hat, for instance, does not prevent contact by a falling object, but it could absorb and/or deflect some of the energy and prevent or minimize injury.
>
> p. 26

Many SMS are focused entirely on the wearing of personal protective equipment, which will deflect an amount of energy but will not prevent the undesired event from happening. Nor will it stop the agency from being in a position to transfer that energy to a person, a piece of equipment, or the environment. Contact control is normally resorted to as an absolute last effort to minimize the degree of energy exchange.

An energy transfer and agency analysis system will assist in directing the actions of a contact control program. Examples of an energy control program are:

1. Prevent the marshalling of energy in the first place. (Prevent workers from climbing to high places from which they may fall.)
2. Reduce the amount of energy that is marshalled from which accidents may result. (Reduce the number of workers permitted to climb to high places.)
3. Prevent the release of energy that has built up. (Build guard rails to prevent falls from high places.)
4. Slow down the release of energy. (Reduce the height from which employees must work.)
5. Separate, in space or time, the energy. (Separate hazardous material from employees.)

SUMMARY

Once a high-risk behavior or a high-risk condition is imminent, there could be an exposure, impact, or exchange of energy. This exchange is in the form of contact with a substance or source of energy greater than the threshold limit of the body or article.

The exposure, impact or contact, or exchange of energy is the portion of the accident sequence that is most closely associated with the loss. The exposure, impact, or exchange of energy is what injures, damages, pollutes, or interrupts the business process.

The energy release theory (Dr. Lestie Ball 1970) states that all accidents are caused by hazards, and all hazards involve energy, either due to involvement with destructive energy sources or due to a lack of critical energy needs.

As part of the measurement of safety performance a monthly and annual progressive analysis of the energy transfer types should be produced. This will indicate trends in which energy transfer types are responsible for the losses.

The agency is the piece of equipment, substance, object, or radiation, most closely associated with the loss. It is the *principal* object such as tools, machine or equipment involved in the exposure, impact, or energy transfer, and is usually the object inflicting injury or property damage. The agency part is that part or area of an agency that inflicted the actual injury or damage.

11 Total Cost of Risk

Although healthy and safe workplace conditions can be justified on a financial basis, many employers prefer to justify them on the basic principle that it is the right thing to do. In discussing safety in industrial and mining operations, it has often been stated that the cost of adequate health and safety measures would be prohibitive and that organizations can't afford it. The answer to that is quite simple and direct: "If you can't afford safety, you can't afford to be in business."

ACCIDENT COSTS

The total economic cost of fatal and nonfatal preventable injury-related incidents in the US during 2020 was $1,158.4 billion. This includes employers' uninsured costs, vehicle damage, fire costs, wage and productivity losses, and medical and administrative expenses. Motor vehicle-related injuries cost $473 billion, work-related injuries cost $163 billion, and public injuries cost $166 billion.

INCREASED PREMIUMS

The end result of an accident can always be translated into costs. Whether the event results in injury, disease, and damage to machinery, property, or materials, or business interruption, they all cost the organization money. Traditionally, these costs have been tolerated as the cost of doing business and have not received management's full attention.

Workers' Compensation normally covers the direct costs of the worker injured in an accident. These premiums can be increased as a result of injury costs, and these increases could remain for three-year cycles or periods. Most of the accidental costs are hidden or indirect costs. These are difficult to calculate and are often ignored because they do not create an immediate financial drain on the organization.

THE EXPERIENCE MODIFICATION RATE (EMR)

The *Experience Modification Rate* (EMR) has a strong impact on an organization. It is a number used by insurance companies to gauge both the past cost of injuries and future chances of risk. The lower the EMR of the business, the lower the worker's compensation insurance premiums will be. An EMR of 1.0 is considered the industry average.

If an organization has an EMR greater than 1.0, the reason for this is that there has been a worker's compensation claim, which the insurance provider has paid. To mitigate the insurance company's risk, they raise the worker's compensation premiums. This increased EMR remains for 3 years.

Workers Comp Premium	Experience Modifier	Modified Premium
$100,000	.80	$80,000
$100,000	1.00	$100,000
$100,000	1.20	$120,000

FIGURE 11.1 A table showing the experience modifier figure in relation to how it affects premiums.

The base premium is calculated by dividing a company's payroll in a given job classification by 100, and then by a *class rate* determined by the National Council on Compensation Insurance (NCCI) that reflects the inherent risk in that job classification. For example, structural ironworkers have an inherently higher risk of injury than shop attendants, so their class rate is significantly higher (Figure 11.1).

INCIDENTAL COSTS

In addition to the obvious costs, there are also the so-called incidental costs of accidents. These incidental costs of accidents have been estimated to be up to five times as great as the actual costs. There have been innumerable studies made, discussions held, articles written, and arguments presented about that figure. No one has disputed the concept that there are indirect, incidental, or hidden costs surrounding accidents. Most tend to agree with that. There is tremendous disagreement, however, about how much those costs might amount to.

HIDDEN ACCIDENT COSTS

The following is a list of 11 types of hidden accident costs, which exclude compensation and liability claims, medical and hospital costs, insurance premiums, and costs of lost time, except when actually paid by the employer without reimbursement. These 11 main hidden costs are:

1. Cost of lost time due to the injured employee.
2. Cost of time lost by foremen, supervisors, or other executives.
3. Cost of time lost by other employees who stopped work.
4. Cost of time spent on the case by first aid attendants and hospital department staff who were not paid for by the insurance carrier.
5. Cost of accident investigation, time, and resources.
6. Cost due to damage to the machine, tools, or other property, or to the spoilage of material.
7. Incidental costs due to interference with production, failure to fill orders on time, loss of bonuses, payment of forfeits, fines, and other similar causes.
8. Cost to the employer in continuing the wages of the injured employee in full.
9. Costs due to the loss of profit on the injured employee's productivity and on idle machines.

10. Costs that occur as a consequence of worker morale due to the accident.
11. Overhead costs per injured employee (light, heat, rent, and other such items, which continue while the injured employee is not productive).

The list of costs as a result of an accident can be a lot longer, depending on the circumstances. Safety pioneers established that there is a definite ratio between the insured (direct) and uninsured (indirect) costs of accidents and *conservatively* put the ratio at 1:5.

TOTALLY HIDDEN COSTS

A further category can be added to the costs of an accident. Those costs are termed the totally hidden costs. These costs are effects that cannot be costed out accurately and include such intangible costs as the cost of pain and suffering of the injured employee, the disruption of the victim's family life, the lowered standard of living due to his or her pay packet being reduced substantially, as well as the accompanying stress and anxiety caused by the accident. The loss to the community and the organization is also difficult to compute.

ICEBERG EFFECT

The iceberg effect is where the majority of accident costs are hidden below the waterline. The hidden costs or indirect costs of accidents include items such as damage, loss of time and production, and business interruptions. Experts have put these costs anywhere from 5 to 50 times more than the direct costs. Irrespective of what the actual figures are, there is definitely a higher cost of hidden costs than direct costs.

MINIMIZING LOSSES

All costs are recuperated from the profits of the company. Peter Drucker, the well-known management consultant and author, said that it would be better to *minimize* losses than to constantly endeavor to *maximize* profits. Management is often aware of efforts to maximize profits by improving quality and productivity but is not always aware of losses occurring as a result of accidents. The prevention of all downgrading accidents is a good investment and can improve the company's bottom line substantially.

In summarizing the advantage of safety control, experts agree that if you save a dollar in accident costs, you add a dollar to profit. All agree that if you add this to the overriding human element, you have the best of both worlds: protection of profits, processes, property, and people. That is why it is so essential to understand and use the accident cause-and-effect sequence (Figure 11.2).

EXAMPLE

In motivating the amount of profit consumed by accident costs, an excellent example is: if an organization's profit margin is 5%, it would have to make sales of $500,000 to pay for $25,000 worth of losses. With a 1% margin, $10,000,000 of sales will be necessary to pay for $100,000 of the costs involved with accidents.

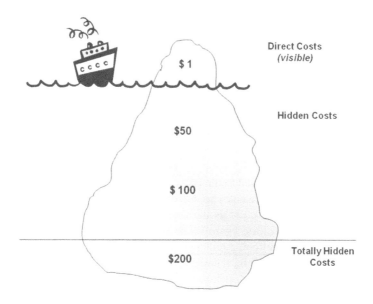

FIGURE 11.2 The iceberg effect which shows more hidden cost below the waterline. (From McKinnon, Ron C. 2012. *Safety Management, Near Miss Identification. Recognition, and Investigation.* CRC Press, Taylor and Francis Group, 6000 broken sound parkway NW, suite 300 Boca Raton, FL. With permission.)

MAIN MOTIVATION

For years, safety practitioners and safety organizations have promoted the humanitarian aspect of safety and endeavored to reduce injury-producing accidents. The humanitarian approach is all very well, but this is not what motivates management.

PROFIT-DRIVEN

Management is in business to make a profit and give shareholders a return on their investment. Profits and the bottom line are what an organization is all about, and what management at all levels is ultimately held accountable for. If the costs of accidents due to poor control can be brought to management's attention, the necessary actions and support will be forthcoming. So many organizations concentrate on the numbers of injuries, and endeavor by all means to reduce these numbers, that they lose sight of their greatest management attention getter, *the total cost of accidents.*

COST REDUCTION

One of the major objectives of accident prevention work is to reduce production or operating costs for the sake of profit. While second to the prevention of human injury, since that is even more directly important as a human value, cost reduction broadens the basis for safety work. Cost reduction provides a direct purpose for preventing all kinds of accidents – accidents that cause injury as well as other undesired

Total Cost of Risk 115

events. Cost reduction brings into focus the losses from property damage and interference with production, as well as those from injury accidents.

Individuals, promoters, and organizations geared to prevent accidental loss will continue to be unsuccessful unless they start to justify their recommendations on a sound financial basis.

During an interview, one executive was asked the question, "How can a company achieve excellence in safety?"; he replied, "Success will not happen overnight. A company must hold safety as a core value right up there with making a profit."

In answering the question, "How are you convincing management to promote safety within their companies," he answered, "If you are trying to convince your company's chief executive officer that safety is important, you must show that safety offers financial rewards. "Cut costs" are two words that corporate America is using today."

In emphasizing the importance of hidden costs, Dan Petersen (1978) says:

> Hidden costs are real. Many people today believe that the dollar is a far better measuring stick than any other in safety, and many companies are beginning to utilize it effectively.
>
> p. 50

It should be remembered that if the severity of injury is a result of Luck Factor 3, then the subsequent direct costs of the injury or disease are also largely fortuitous, and therefore, although the dollar cost is a useful measurement, it, like injury rates, is highly subjective.

FINES

Most organizations are required by legislation to maintain a safety management system (SMS), which gives certain controls over risks that have been identified. Although these controls, checks, and balances do cost money, they should be viewed as an investment, as accidental losses could be far more expensive. Non-compliance with health and safety legislation could also cost a company a staggering amount of money.

COST OF NON-COMPLIANCE

The cost of non-compliance with health and safety laws and regulations can cost an organization a great deal, even before the occurrence of an accident. The Occupational Safety and Health Administration (OSHA) (2023) lists the top 10 violations of 2021 and the fines (rounded) for non-compliance.

1. Fall Protection, construction – penalties: US$29 million.
2. Respiratory Protection – penalties: US$4 million.
3. Ladders in Construction – penalties: US$5 million.
4. Hazard Communication – penalties: US$3 million.
5. Scaffolding in Construction – penalties: US$6 million.

6. Fall Protection – Training – penalties: US$3 million.
7. Hazardous Energy (Lockout/Tagout) – penalties: US$12 million.
8. Eye and Face Protection – penalties: US$4 million.
9. Powered Industrial Trucks – penalties: US$5 million.
10. Machinery and Machine Guarding – penalties: US$10 million (OSHA Website 2023).

Increased Fines and Penalties

For the year 2023, OSHA's maximum penalties for serious and other-than-serious violations will increase from US$14,502 per violation to US$15,625 per violation. The maximum penalty for willful or repeated violations will increase from US$145,027 per violation to US$156,259 per violation.

In 2021, a committee in Congress proposed maximum penalties of $70,000 for a "serious" violation and $700,000 for a "willful" or "repeat" violation. The recommendation, however, failed to make its way into the 2022 Inflation Reduction Act.

HIGHEST EVER

The US Mine Safety and Health Administration (MSHA) (2010) issued the largest fine in its history, US$10.8 billion against the former Massey Energy Co. in connection with the 2010 Upper Big Branch Mine explosion. In its December 6 fatal accident investigation report, MSHA attributed the root cause of the disaster to a corporate culture that valued production over safety.

On April 5, 2010, a massive explosion in the Upper Big Branch Mine, which was operated by Performance Coal Co. (PCC), a subsidiary of Massey Energy Co., killed 29 miners and injured two others. MSHA has now issued Massey and PCC 369 citations and orders, including an unprecedented 21 flagrant violations, which carry the most serious civil penalties available under the law.

"The results of the investigation led to the conclusion that PCC/Massey promoted and enforced a workplace culture that valued production over safety and broke the law as they endangered the lives of their miners," said Secretary of Labor Hilda L. Solis. "By issuing the largest fine in MSHA's history, I hope to send a strong message that the safety of miners must come first."

MSHA announced its report findings and fines following the US$209 million settlement and non-prosecution agreement reached Dec. 6 among the U.S. Attorney's Office for the Southern District of West Virginia, the U.S. Department of Justice, Alpha Natural Resources Inc., and Alpha Appalachia Holdings Inc., formerly known as Massey Energy Co. (MSHA website 2023).

MAJOR ACCIDENTS

The Chernobyl nuclear accident cost around US$200 billion. The loss of the space shuttle Columbia cost US$18 million in direct costs to the accident investigation board and US$112 million in support of the board's investigation. The Prestige oil spill cost US$12 billion. The oil spill resulting from the Exxon Valdez disaster cost

US$2.5 billion, and that of the Deepwater Horizon cost US$61.6 billion. The Bhopal accident cost US$470 million in settlements alone.

COST–BENEFIT

An effective approach to the safety problem, which is supported by many other practitioners, is to do a cost–benefit analysis during the evaluation phase of a risk assessment. A thorough risk assessment will indicate to management where best to spend its money for the largest return. Traditionally, a number of safety systems have channeled money into activities that have not necessarily tackled the root cause of accident problems.

When various risk reduction approaches are being considered, one basis for decision-making is to examine the cost of each versus the benefit of its application. The alternate measures that produce the greatest benefits for the least cost are normally selected. One problem with this approach is the difficulty in assessing the dollar costs of such intangible accident consequences as pain, suffering, and loss of life.

LIFE's VALUE

In 2020, it was estimated that the cost of a US work-related fatality was US$1,290,000 and a disabling injury was between US$37,000 and US$43,000, depending on employer costs.

One of the largest penalties paid by a South African division of an international organization was when, in 1994, four chemical firms agreed to pay 20 South African workers ZAR 9,4 million (US$1.5 million) in damages and costs.

REPUTATION

One of the hidden costs of an accident is the cost a company suffers when its reputation is tarnished as a result of a serious accident or series of accidents. On the day following the Vaal Reefs disaster in South Africa on 17 May 1995, the shares of Anglo-American Gold Division, the owners of Vaal Reefs, fell by ZAR16 (US$3) per share in one day.

SEVERE REPERCUSSIONS

Even ignoring the intangible costs, an accident that results in a fatality normally has severe repercussions for the organization. In one particular fatal investigation, the costs were broken up into injury, property damage, business interruption, and total hidden costs. The summary of this accident report was, "The losses of the accident are listed under the headings of injury, property damage, interruptions, and total hidden costs. Total cost is estimated to be far in excess of US$1,000,000."

Even property-damage accidents can hamper the production of an organization, and one particular property-damage accident, which resulted in a seven-day period to repair the damage, resulted in a loss of 5,000 units being produced.

WHITE PAPER

A survey was conducted that asked management to list the most important contribution health and safety personnel can make to their companies. Only 37% of the managers listed "document the financial impacts of safety activity." When asked what skills and knowledge are important for a safety professional, 17% responded, "The ability to document dollar savings of safety activity."

TOTAL COST OF RISK

There are three main areas where costs play a role in safety. The first is the cost of the end result of undesired events such as injury and property-damage accidents. As discussed, these costs could have as many as three different tiers and can range from tangible to intangible and from direct to indirect costs. They are losses to the organization, nevertheless. The second cost of risks is the cost to insure equipment, plants, products, and personnel. These are the worker compensation costs, insurance costs, and such like. The third cost is involved in reducing, containing, and minimizing the risks that could manifest in losses and cost-producing events. This is the cost of the health and SMS and personnel.

SUMMARY

The end result of an accident can always be translated into costs. Whether the event results in injury, disease, damage to machinery, property, or materials, or business interruption, they all cost the organization money. Traditionally, these costs have been tolerated as the cost of doing business and have not received management's full attention.

The accident sequence always ends up with costs as the last effect. Once the event results in exposure, impact, or contact with a source of energy, the losses could be due to injury, property damage, business interruption, or a combination of all three. The costs of assessing and controlling the risk have proved to be less expensive than the cost of the consequence of the event. The benefit of reducing and controlling the risk is also the avoidance of heavy fines for non-compliance with legal safety standards. The costs of safety controls are therefore a good investment, and as one safety professional put it, "If you think safety is expensive, try an accident!"

Part IV

Leading Safety Management
Performance Indicators

12 Safety Management Control

The management control function has eight steps:

I – Identify hazards and assess the risk.
I – Identify the work to be done to control the risks.
S – Set standards of measurement.
S – Set standards of accountability.
M – Measure conformance to standards by inspection.
E – Evaluate conformances and achievements.
C – Correct deviations from standards.
C – Commend compliance.

LEADING SAFETY MANAGEMENT PERFORMANCE INDICATORS (MEASUREMENTS OF CONTROL)

Leading or proactive safety performance indicators are measurements that record actions, activities, processes, and controls before the occurrence of accidental loss.

DEFINITION

Leading safety performance indicators are measurements of management control. They are positive performance indicators and are proactive, leading, metrics that drive safety performance.

SAFETY CONTROLLING

Safety controlling is the management function of identifying what must be done for health and safety, setting health and safety standards, inspecting to verify completion of work, evaluating, and following up with safety action. This is the most important safety management function and is vital to prevent downgrading events in the form of accidental injury, damage, or business interruption.

RISK-BASED, MANAGEMENT-LED, AUDIT-DRIVEN SAFETY MANAGEMENT SYSTEM (SMS)

Based on risk assessments, a manager determines and schedules the work needed to be done to create a healthy and safe work environment and to eliminate high-risk behaviors of employees and high-risk workplace conditions. This would mean the introduction of a suitable risk-based, management-led, and audit-driven structured health and safety management system (SMS) based on world's best practices and aligned to the risks of the organization. The SMS is driven by health and safety standards, which are measurable management performance criteria. Each standard must set levels of performance and conformance that can be measured at regular intervals. All SMS should be based on the nature of the business and be risk-based, management-led, and audit-driven. The management control function has eight steps:

I – Identify hazards and assess the risk.
I – Identify the work to be done to control the risks.
S – Set standards of measurement.
S – Set standards of accountability.
M – Measure conformance to standards by inspection.
E – Evaluate conformances and achievements.
C – Correct deviations from standards.
C – Commend compliance.

Step 1 – Identify the Hazards and Assess the Risk

The hazard identification and risk assessment (HIRA) process will ensure that an undertaking has identified all the hazards, analyzed the risks, evaluated them, and ascertained which risk control methods to apply. These controls would form the basis of the SMS. The frequency and quality of the inspections and hazard control actions are measurable, leading metrics of safety performance.

Step 2 – Identify the Work to be Done to Control and Mitigate Risks

Once the risks have been assessed, evaluated, and prioritized, it is now management's function to identify what work must be done to ensure the treatment of the risks. The risk assessment would have identified both physical and behavioral risks. Management can now implement certain control elements under the umbrella of the SMS to reduce the risk as much as is reasonably practical. The risks identified can be measured and their reduction monitored as a safety metric.

Examples of the work that may need to be done are the following:

- Controlling permit required work.
- Guarding of all machinery and pinch points.
- Regular inspections of lifting gear.
- Hazard identification and risk assessments.
- Providing and maintaining personal protective equipment (PPE).
- Formal accident investigation procedures.
- Hazardous substance control.
- Legal injury and disease reporting.
- Safety induction training for new employees.
- Hazardous work procedures and controls.
- Establishing policies and standards, etc.

MEASURABLE PERFORMANCE STANDARDS

The above are a few components of a structured SMS and are driven by standards which can be measured and quantified. These are leading, positive indicators of safety performance.

Step 3 – Set Standards of Measurement

Managers get what they want, and once management sets health and safety standards, these standards are usually achieved by the organization. Setting standards of measurement clearly indicates how things must be in the work environment. Health and safety standards are measurable management performances. Standards must be in writing and should contain the following headings:

- Title
- Purpose.
- Resources needed.
- Authority, responsibility, and accountability.
- Legal requirements.
- General requirements.
- Monitoring of activities.
- Documentation control.

By setting standards of measurement, management defines the direction in which the organization moves. Should management set a standard for good housekeeping practices, this standard, a measurable management performance, will then dictate how housekeeping is managed in the future. What gets measured gets done. Standards give the company goals and directions and a definite focus on the end result. These standards can be of *measurement* and of *performance* and ask the questions, "What must the end result be?" and "What must be done, by whom, and by when?"

All SMS standards should be SMART, meaning that they should be:

- Specific – The standard must specify exactly what must be done in detail. It should not be vague or generalize.
- Measurable and manageable – The standard must be measurable and manageable.
- Achievable and advantageous – The standard must be achievable considering costs and resources. It should be aligned with the organization's objectives and be advantageous to the organization.
- Realistic and result-oriented – Standards must be realistic and result-orientated. An indicator such as "injury-free" sounds nice, is ideal, but is simply not realistic in a workplace setting. A measurable performance standard, such as the holding of safety committee meetings monthly, is achievable and realistic.
- Time-bound – the standard must specify when and how often the actions prescribed must be carried out.

Standards of measurement should be set for all the elements, programs, and processes within the SMS. A comprehensive SMS would contain about 80 elements, programs, and processes. They would include:

- Housekeeping.
- Stacking and storage.
- Hygiene monitoring.
- Environmental conformance.
- Safety committees.
- Health and safety training.
- Safe work procedures.
- Risk assessments.
- Plant inspections, etc.

Step 4 – Set Standards of Accountability

The next step in the control process is the setting of standards of accountability. A standard of accountability indicates *who* will do *what* and by *when*. Setting standards of accountability asks, "Who must do it and by when?" An example of setting standards of accountability is the follow up action after an accident investigation. The control steps to prevent a recurrence are what need to be done to prevent a recurrence of the accident. This should be followed by making people or departments responsible for the action, as well as committing those people or departments to a date for completion.

There is often confusion about *authority*, *responsibility*, and *accountability* and it is opportune to define these concepts.

SAFETY AUTHORITY

Safety authority is the total influence, rights, and ability of the position, to command and demand safety. Management has ultimate safety authority; therefore, it is the only level that can effectively implement and maintain an effective SMS. Leadership has the authority to demand the implementation of safety system elements and also the authority to take necessary remedial actions to ensure that standards, policies, and procedures are implemented and maintained.

SAFETY RESPONSIBILITY

Safety management system guidelines such as the International Organization for Standardization (ISO) ISO 45001-2018, *Occupational Health and Safety Management Systems*, the Occupational Safety and Health Administration (OSHA) *Voluntary Protection Program* (VPP), and others emphasize that occupational safety and health are the responsibility of management, starting with the most senior managers. This cannot be overemphasized, as many still think that safety is the sole responsibility of the safety manager and the safety department. This is one of the biggest safety paradigms that needs to be changed.

Safety responsibility is the safety function allocated to a post. It is the duty and function demanded by the position within the organization. This duty lies with all

Safety Management Control

levels of management as well as with employees. The higher the management position, the higher the degree of safety authority, responsibility, and accountability. One cannot be held accountable for something over which one has no authority. The degree of safety accountability is proportionate to the degree of safety authority. Job descriptions are vital management tools and should clearly define the safety authority, responsibility, and accountability for all levels within the organization. The safety management system's safety standard must clearly define these relationships for the system to be a success.

SAFETY ACCOUNTABILITY

Safety accountability is when a manager is under an obligation to ensure that safety responsibility and authority are used to achieve both SMS and legal safety standards. Employees also have safety accountabilities, but in proportion to their safety authority.

Leadership has the accountability to manage the SMS and its components and to provide the necessary infrastructure and training to enable the system to work. Employees should be held accountable for participating in the system and following company safety policies, procedures, and practices.

Management at all levels is then held accountable to rectify the problems identified by the ongoing risk assessment process, as well as the management review and safety system audits, to ensure that the high-risk acts or conditions highlighted by these systems are rectified and do not recur.

Setting safety standards of responsibility involves deciding *who* will do *what* and *when*. An example of a standard of responsibility is the role played by health and safety representatives. Health and safety representatives have been given the authority to inspect their immediate work area using a safety element checklist as a guideline. This inspection is carried out monthly as prescribed by the standard. The standard of responsibility is:

1. Who? – The appointed health and safety representative.
2. Will do what? – Will carry out an inspection of his/her work area.
3. When? – This inspection will be carried out on a monthly basis.

One of the many safety myths is that "everybody" is responsible for safety. Individuals can only be *responsible* for items and people over whom they have *authority* and can thus be held *accountable* for only those conditions and people over whom they have authority.

Step 5 – Measurement Against the Standard

This control function is when management measures what is actually happening in the workplace against the preset standards. To measure successfully, a walkabout inspection and control documentation review must be carried out. Employees doing these inspections should be aware of and familiar with the standards. One of the greatest failings in most SMS is insufficient or inadequate inspections.

An inspection is not an audit. An audit is not an inspection. An audit inspection forms part of an audit. The two concepts should not be confused.

System Standards

Systems to enable ongoing measurement against standards are part of a SMS, and these could include the monthly inspections of local work areas by appointed health and safety representatives. Safety inspections are ideal measurement tools. Critical task observations also allow opportunity for measurement against standards. The setting of standards and constant measuring against those standards immediately identify the strengths and weaknesses of the SMS. Safety personnel should also conduct formal inspections on a regular basis and compare actual SMS processes and procedures with the standards. A checklist should always be used when doing these inspections, as it will serve as a constant reminder of what must be measured.

Measuring Performance

Safety management control and the measurement of the control process are not merely measuring and comparing injury statistics with other companies or industries. This is a management function that measures whether the organization is living up to the norms agreed to by management and employees in the form of SMS health and safety standards and the health and safety policy statement. Each element of the SMS has a protocol against which it is measured. Points are allocated to every standard, minimum standard, and minimum standard detail of the element to form an audit protocol.

Measurements include reports of downgrading incidents, deviations from standards, and input and output metrics.

Critical Task Observation

Another form of measurement against standards is critical task observation. This involves observing an employee carrying out a critical task while following the steps of the written safe work procedure. The written safe work procedure is the end result of critical task identification (task risk assessment) and the critical task analysis process. The observation allows for the measurement of workers' performance during the critical task against the prescribed performance dictated by the procedure. The procedure sets a standard of performance.

Step 6 – Evaluate Conformances to SMS Standards and Achievements

The evaluation process is the quantification of the degree of conformance to the standards established. The legal health and safety requirements are viewed as the minimum standard to achieve. Evaluation of the achievement of standards is normally facilitated through the safety management system audit process. An SMS should be driven by SMS audits. These regular audits systematically quantify the degree of compliance with standards. They evaluate the management work being done to combat losses. What gets measured gets attention, and consequently, the evaluation of compliance with safety standards gives an indication of what is being done and what is not being done. The quantification of safety control actions is far more reliable and significant than the measurement of safety consequences, which are largely fortuitous.

Safety Management Control 127

External Audits

A number of health, safety, and risk-management organizations provide auditing services for clients. This external audit is of tremendous value to any organization as it is totally impartial and conducted by auditors who are thoroughly familiar with the company's audit protocol. It is again emphasized that a safety inspection is not an audit. The audit of an organization's safety system is a structured approach to the quantification of safety compliance and adheres to the following sequence:

1. Pre-audit meeting.
2. Audit facilities.
3. Audit team.
4. Physical inspection.
5. Compliance audit.
6. Systems audit.
7. Documentation review.
8. Verification of the disabling injury incidence rate.
9. Management close-out and audit results.
10. Audit report.

Retrospective

The audit results and percentage achievement are normally based on the SMS achievements during the preceding 12 months. Credit is not given for good intentions but rather for programs, processes, and procedures that have been in operation for at least six months.

Internal Audit

Internal audits of the entire SMS to evaluate conformance with standards should be carried out every six months. The internal audit system should follow the same guidelines as an external audit and will culminate in a percentage of conformance as well as a breakdown of conformance against standards for each element.

Figure 12.1 shows the SMS section elements 5.1–5.23 with the percentage conformance for Year 1 and Year 2, evaluated during the audits. This immediately gives management an indication as to the strengths and weaknesses of the SMS processes. It also indicates where conformance and non-conformance to health and safety standards lie.

Each element of the SMS that does not score 100% indicates a weakness in the element, indicating that the standards established for that element have not been fully met. This indicates to management which elements of the SMS require action and are positive indicators of safety management performance.

Weighting Systems

Although traditional audit score weighting systems allocate different points for each SMS element's minimum standard detail, the weighting may not be relevant in certain industries and, under certain circumstances, could weaken the intent of the audit process. To weigh each element – minimum standard and minimum standard detail

5.1	Chief executive officer responsible for health and safety	5.13	Induction and job health & safety training
5.2	Person(s) made responsible for occupational health and safety co-ordination	5.14	Formal health and safety training annually
5.3	Appointment of health and safety representatives	5.15	Medical examinations
5.4	Health & safety committees		Selection and placement
5.5	On-the-job health and safety communication	5.16	Plant inspection – safety representatives
5.6	First-aid and facilities	5.17	Bi-annual internal self-audits
5.7	First-aid training	5.18	Health and safety design specifications, etc.
5.8	Posters, bulletins, newsletters, health & safety videos, internal competitions	5.19	Written safe work procedures: issued and used
5.9	Injury and diseases experience and star grading board	5.20	Planned task observations
5.10	Suggestion scheme	5.21	Work permits
5.11	Health and safety reference library	5.22	Off-the-job safety
5.12	Annual report – monthly, quarterly, or bi-annually	5.23	Health and safety policy

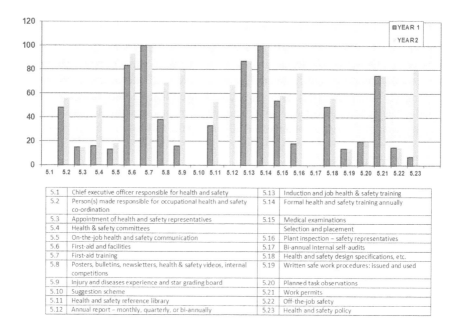

FIGURE 12.1 An example of part of an internal audit report showing the elements, the element number, and the percentage scores for Year 1 and Year 2.

equally – would be an ideal situation. Some SMSs have incorporated a 0–5 rating (Figure 12.2) for each minimum standard detail of the SMS. This rating implies that each minimum standard detail is as important as the next. The scale indicates 0% compliance to 100% compliance.

Step 7 – Correct Deviations from Standards

Corrective action is the safety management work that must be done to correct those activities that were not completely controlled. If any critical SMS element is evaluated at less than 100%, some action needs to be taken to ensure total conformance with standards. Management must do what it says it is going to do. The safety standards indicate what must be done. Deviations indicate that the safety objective has not yet been achieved.

Accident Investigation

An accident is caused by a failure in the management system, and after an investigation, certain action plans or controls are recommended to prevent the recurrence of a

2.7 Hazardous Substances Control (15)	0	1	2	3	4	5	Hazard	Notes	Action	Date
Alphabetical list of all substances										
Person designated to coordinate the list of MSDS										
MSDS sheets available and used in departments										

FIGURE 12.2 An example of a 0–5 scoring system for an element of a health and safety management system (SMS).

similar accident. This is corrective action and should be directed at the root causes of the problem, not merely the symptoms. Correcting high-risk behavior and high-risk conditions may provide temporary relief, but the real cause must be identified and the problem solved.

Step 8 – Commendation for Compliance

One of the main failings in numerous safety processes and programs is the lack of commendation and recognition. Commendation should be given for the achievement of objectives. If a department meets and maintains the housekeeping standard, for example, the entire group should be commended. Commendation for pre-contact safety activities is far more effective than commendation for injury-free periods.

COMMENDATION

If, as a result of the measurement and evaluation, a high degree of conformance to performance standards is found, commendation should be given. Safety, as a profession, has often been guilty of emphasizing the lack of control and not complimenting where good control exists.

RECOGNITION FOR THE ACHIEVEMENT OF PROACTIVE OBJECTIVES

People at workplaces and in other walks of life thrive on recognition and acknowledgment. Recognizing and acknowledging people for their safety work should be done as often as possible. Traditionally, safety recognition was only given to individuals for being "injury-free" or for having worked a certain number of days without a lost-time injury. That could just be the result of good luck.

Maintaining good housekeeping and carrying out monthly inspections of ladders, portable electrical equipment, hand tools, personal protective equipment, etc. is an ongoing control system, and this effort should be recognized.

MANAGEMENT

It is good management practice to commend employees for their safety efforts. Bearing in mind that control is pre-contact accident control, this is more important than the recognition of no adverse consequence in the form of severe injury.

CONCLUSION

Only management has the authority to create a healthy and safe workplace. By implementing measurable management control measures in the form of a risk-based, management-led, and audit-driven SMS, hazards and their associated risks will be identified and reduced, making the workplace safer for all.

SUMMARY

Management control is the most important safety management function and is vital to prevent downgrading events in the form of accidental injury, damage, or business interruption.

Leading safety performance indicators are measurements of management control. They are positive performance indicators and are proactive, leading metrics that drive safety performance.

Based on risk assessments, a manager determines and schedules the work needed to be done to create a healthy and safe work environment and to eliminate high-risk behavior of employees and high-risk workplace conditions. This would mean the introduction of a suitable risk-based, management-led, and audit-driven structured health and safety management system (SMS) based on world's best practices and aligned to the risks of the organization.

The management control function has eight steps:

I – Identify hazards and assess the risk.
I – Identify the work to be done to control the risks.
S – Set standards of measurement.
S – Set standards of accountability.
M – Measure conformance to standards by inspection.
E – Evaluate conformances and achievements.
C – Correct deviations from standards.
C – Commend compliance.

Part V

Examples of Positive Performance Safety Indicators (PPSIs)

13 Near Miss Incidents as a Measurement of Safety Performance

SIGNIFICANCE OF NEAR MISS INCIDENT RECORDING AND MEASUREMENT

The recording and investigation of near miss incidents is one of the best measurements of safety management performance. Reported near miss incidents show the base of the iceberg, where the real safety problems lie. This measurement is predictive, as near miss incidents are accident precursors. Near miss records tell an organization where they are going, while accidents tell an organization where they have been.

NOT JUST QUANTITY

While the number of near miss incidents reported is a positive leading measurement of safety performance, the number of near miss incidents that have been risk ranked and actioned for rectification is just as important. Feedback submitted to the reporters is also important.

LEADING PERFORMANCE INDICATORS

A near miss incident is an undesired event that, under slightly different circumstances, could have resulted in injury, property damage, or business interruption. Near miss incidents with a higher potential for accidental loss are termed HIPO (high-potential near miss incidents).

Many argue that a near miss incident is both a lagging and leading indicator because the undesired event (inadvertent flow of energy) occurred but no loss (no energy transfer) was experienced, and since no loss occurred, it must be classified as a leading indicator. Action can be taken before a loss occurs in the future, which means it is a lead indicator. As Dan Petersen (1998) said about near miss incidents,

> A firm can dictate, in advance, what actions it should take to prevent accidents, and then it can measure how well these predetermined actions are executed.
>
> **p. 37**

A near miss incident is a lagging indicator since the undesired event has already occurred. This means a high-risk situation existed that led to an unintended flow of energy. The flow of energy did not result in an exposure, impact, or transfer of energy purely due to fortuity. This fortuity was due to slightly different circumstances or positioning or timing. Had someone been at the right place at the right time, a loss would have occurred, making the near miss incident a potential lagging indicator. However, since there was no injury, it is also viewed as a leading indicator of safety management performance.

NEAR MISS INCIDENTS (CLOSE CALLS)

Near miss incidents have also been referred to as *close calls, near hits, and narrow escapes*. To clarify the concept, the following definitions are given:

- *An undesired event, which, under slightly different circumstances, could have resulted in harm to people, property damage, or business disruption.*
- *An accident with no injury or loss.*
- *An event that narrowly missed causing injury or damage.*
- *An incident where, given a slight shift in time or distance, injury, ill-health, or damage easily could have occurred, but didn't this time around.*

ICEBERG EFFECT

Near miss incidents lurk below the waterline and are the accidents an organization has not yet had. Identifying, reporting, and rectifying the root causes of these events will lead to the prevention of future events that could have possible devastating consequences. Near miss incidents are warnings as they are accident precursors.

THE ACCIDENT RATIO

In revisiting the example accident ratio discussed in Chapter 7, the analysis of the accident ratio produced (Figure 13.1) shows a high number of minor injuries per serious injury. It shows the correlation between property damage accidents and serious

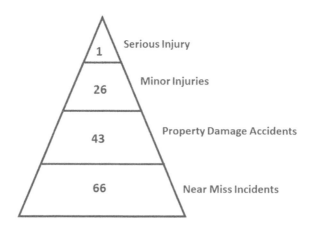

FIGURE 13.1 The organization's derived (example) accident ratio shows 66 near miss incidents reported for every serious injury recorded.

injuries as well as a low report of near miss incidents, although reporting is occurring. The goal of the accident ratio is to show that the number of near miss incidents will be reduced by effective safety management controls, which in turn will reduce property damage accidents, minor injuries, and serious injuries. The base of the triangle must be tackled first before injuries and damage events can be reduced.

LARGE NUMBERS

Regrettably, in the beginning stages of near miss incident reporting, a large number will be reported, which should be viewed as a positive sign rather than a negative one. In the case of near miss incident reporting, the more incidents reported, the better the opportunities to take action to prevent recurrences.

Lagging safety measurements are injuries and property damage events. Leading indicators are the near miss incidents that have not resulted in losses.

BENEFITS OF NEAR MISS INCIDENT REPORTING

The benefits of reporting and acting on near miss incidents are as follows:

- Action can be taken before an accident occurs.
- Weaknesses in the safety management system are revealed before a loss occurs.
- Each near miss incident whose root cause is eliminated is one less potential accident.
- It involves employees reporting accident precursors.
- Employees are informed of the corrective actions taken.
- It directs action to where the problems lie.
- It is a predictive technique, not a reactive one.

MEASUREMENT OF NEAR MISS INCIDENT REPORTING

The reporting of near miss incidents is the first step in the near miss system and should not be the only gauge used for measurement.

The near miss incident system should not restrict reporters to reporting only near miss incidents. As they are so closely related, reports of high-risk work environments and high-risk practices should also be encouraged. Many near miss reports will in fact be high-risk behavior or conditions, but the reporters should not be discouraged from reporting if they are not pure near miss incidents as per definition. Any hazard reported is an opportunity to prevent loss.

FORMAL REPORTING

A formal reporting system is where the observer fills out the report form (or electronic report) and posts it in the near miss box or hands it in to the supervisor, who then enters it into the system. The formal system is the main reporting method but requires access to a form, or a computer terminal, or a safety reporting hotline.

NEAR MISS INCIDENT REPORT
Name: (Optional)
Date: Location:
Near miss Incident:
[risk matrix: Probability of Occurrence (Low, Medium, Medium-high, High) vs Potential Severity (Low, Medium, Medium-high, High)]
Describe What Happened:
Action Taken:

FIGURE 13.2 A near miss incident report form.

The near miss incident report form (Figure 13.2) can be printed in pocket-size booklets so that they can easily be carried by employees who can complete the form immediately upon the occurrence of a near miss incident.

NEAR MISS INCIDENT REPORT RECORD LOG

Figure 13.3 shows an example of a near miss incident record log. These are actual reports from a work area. The near miss incident is allocated a tracking number, and the potential severity (S) and frequency of recurrence (F) are recorded from the risk matrix on the report form, as well as the potential energy exchange that could have occurred under different circumstances.

Number	Date	Area	Near Miss Incident	Severity	Frequency	Energy Exchange
14	2012/03/23	Maint	While oiling the pulley shaft the machine guard fell off almost falling on worker's foot.	M	L	Hit by
15	2012/03/25		A fork-lift truck was speeding and had to swerve to avoid hitting a worker.	S	L	Struck by
16	2012/04/01		A tool fell from the contractor's scaffold. Men were walking past and were almost struck by the falling object.	M	M	Falling object

FIGURE 13.3 An example of a near miss incident report record log.

Hazard Reporting System

Some organizations incorporate a hazard reporting system into their safety management system. This is a system within the safety management system (SMS) whereby employees are encouraged to report high-risk workplace conditions, high-risk behaviors, near miss incidents, and other downgrading events (Figure 13.4).

The system incorporates a reporting form (or electronic reporting system) and drop-off points where the forms can be deposited. As with the near miss incident reporting system, action should be taken on items reported, and feedback should be given. To complement a successful near miss incident reporting system, this hazard reporting system should not necessarily require the reporter's name.

These hazard reports are leading indicators, as they are actions taken to identify system failures before an accident occurs. The monthly and annual metrics could include the following:

- Total number of hazards reported.
- Number of high-risk workplace conditions noted.

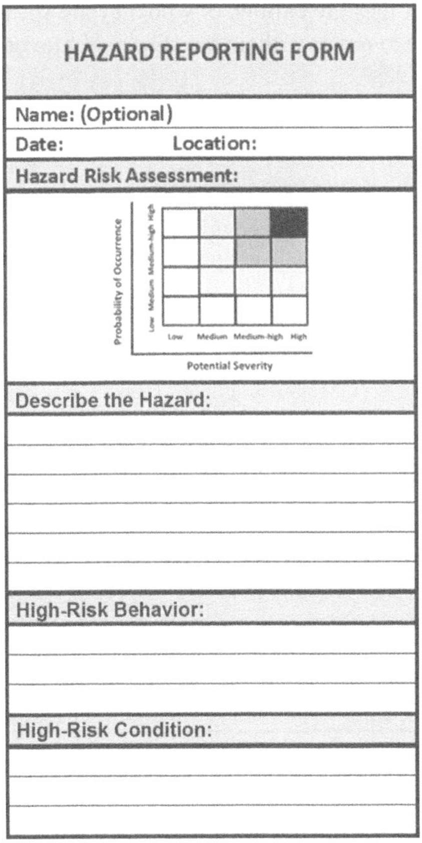

HAZARD REPORTING FORM

Name: (Optional)

Date: Location:

Hazard Risk Assessment:

Describe the Hazard:

High-Risk Behavior:

High-Risk Condition:

FIGURE 13.4 An example of a hazard reporting form.

- Number of high-risk behaviors noted.
- Number of hazards rectified.

5-POINT CHECKLIST

The 5-point safety checklist (Figure 13.5) is used in many mines and fulfills the functions of a safety observation program. The form is foldable and pocket-sized. Each miner is required to complete and submit a 5-point checklist after every work shift. The example in Figure 13.5 is a checklist which is used and incorporates the reporting of accidents, injuries, and near miss incidents.

SAFETY REPORTING HOTLINE

To encourage the reporting of high-potential near misses and other safety issues that require urgent attention, the installation of a dedicated internal telephone number as a safety hotline is advised. This means any employee can use the internal phone system to report near miss incidents or other safety-related matters. The caller remains anonymous and does not have to identify themselves. No one in the immediate vicinity knows what number they are calling, or who they are speaking to, and this gives the reporter confidence to report without the chance of him or her being ridiculed or teased by other colleagues.

INFORMAL REPORTING

Informal reporting is when near miss incidents are reported verbally without filling out a form or online report. This could be an employee who tells his supervisor

5-POINT SAFETY SYSTEM CHECKLIST	Did you find an unsafe condition today?
☐ PPE – Do you have required and necessary Personal Protective Equipment?	If so, please note below:
☐ Permits/Procedures Followed: Lockout/Tagout, Burning, Confined Space	**Did you have an accident, near miss or injury today?** ☐ YES ☐ NO
☐ Pre-use inspection completed	If so, please note below:
☐ Defects noted in logbook	
☐ Equipment is clean	
☐ Chock blocks available and in use	**Environment, Health, Safety, Comments:**
☐ Berms meet standards	
☐ Roadways meet standard	
☐ Lighting meets standard	
☐ Signage meets standard	
☐ Housekeeping meets standard	

FIGURE 13.5 An example of a 5-point checklist.

about a safety deviation. It could be an employee telling another employee about a certain situation. Observations discussed at safety meetings and tailgate safety sessions would fall into the informal category.

All safety meetings and pre-task talks should begin with an incident recall session as part of the informal reporting process. The discussions from incident and accident recall sessions are informal reports and form an invaluable part of the reporting system.

RISK RANKING

The risk ranking of the event reported may deter reporters from reporting and participating in the system if the procedure is too complex and they do not understand it fully. The best is to keep it very simple. During training sessions on near miss incident reporting, the attendees should go through the process of actually ranking examples of near miss incidents so that they are familiar with the process. They should be taught how to use the simple risk matrix, which should be incorporated into the report form, and should understand that risk assessment is not a perfect science and that we all see risk differently. There should be no concern if different assessors give different rankings for the same hazard. People see things differently and have different perceptions of risk.

Figure 13.6 shows a simple risk matrix that allows the reporter to rank the potential recurrence and severity of the event being reported. Near miss incidents ranked in the high and medium-high ranges should receive priority action over the lower ranked risks.

MEASUREMENT CRITERIA

The following are metrics that could be used to measure safety performance related to near miss incidents.

- Total near miss incidents reported in the period.
- Total near miss incidents that have been risk-ranked.

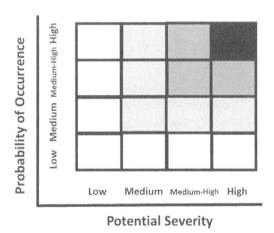

FIGURE 13.6 A simple method of ranking near miss incidents using a risk matrix.

- Action initiated to eliminate the near miss incident's root causes.
- Completed actions to eliminate the near miss incident's root causes.
- Feedback information on receipt and progress of near miss incident's remedial actions.

CALCULATIONS

The following formulas will assist in compiling meaningful near miss incident statistics:

Number of near miss incidents reported per 100 employees:

$$\% \text{ Reporting} = \frac{\text{Reports received}}{\text{Total employees}} \times 100$$

Risk ranking of near miss incident completed:

$$\% \text{ Ranking completed} = \frac{\text{Reports ranked}}{\text{Total reports}} \times 100$$

Action taken on the near miss incidents:

$$\% \text{ Actioned} = \frac{\text{Reports actioned}}{\text{Total reports}} \times 100$$

Actions completed:

$$\% \text{ Actions completed} = \frac{\text{Actions completed}}{\text{Total reports}} \times 100$$

Feedback submitted:

$$\% \text{ Feedback sent} = \frac{\text{Report feedback sent}}{\text{Total reports}} \times 100$$

MONTHLY TOTALS

The monthly average of the five criteria can be tabulated in a table (Figure 13.7) and used as a leading indicator of safety management performance.

RECTIFICATION SYSTEM

Each reported near miss incident should be allocated a tracking number, and this number should be used to track the near miss incident until all the corrective actions have been completed.

Near Miss Incidents as a Measurement of Safety Performance 141

MERTIC	ACHIEVEMENT	SCORE	TARGET
Reports Received	Per 100 Workers	%	50%
Risk Ranking Completed	Near Miss Incidents Ranked	%	100%
Action Taken	Action Taken	%	100%
Actions Completed	Actions Completed	%	100%
Feedback	Feedback Circulated	%	100%
MONTH – April	AVERAGE TOTAL	%	

FIGURE 13.7 A monthly progress table to record near miss reports and corrective actions.

Feedback

One of the biggest failings of a near miss incident reporting system is that feedback on reported near miss incidents is not forthcoming. Employees who report a near miss incident need reassurance that management is taking action to rectify the causes of the incident, so feedback on progress is important. This feedback could be given in the safety newsletter, on bulletin boards, or discussed as an agenda item at safety committee meetings.

Setting Targets

Allocating targets for the number of near miss incident reports to supervisors, divisions, or individual workers is not always the best method to encourage reporting. Such targets could lead to false reporting to meet the target figures. Targets should be set for the organization as a whole. This target should be reasonable, realistic, obtainable, and a good kick-off target of 50% of the workforce reporting near miss incidents is a good start. Once employees understand what a near miss incident is and how to report it without getting into any kind of trouble, reporting will increase. Insisting on a number of reports per supervisor or division opens the system up to manipulation.

EXAMPLE METRICS

Monthly reports on all aspects of the near miss incident reporting system should be produced in graphic format (Figure 13.8) circulated and discussed at safety committee meetings.

Outstanding actions can also be monitored on a monthly basis and reported on at safety committee meetings (Figure 13.9).

PROACTIVE SAFETY PERFORMANCE MEASUREMENT

The reporting, risk ranking, and rectification of the root causes of near miss incidents is one of the most proactive safety management actions, and these records are positive measures of good safety performance as they give a clear picture of system failures and what could have happened under different circumstances.

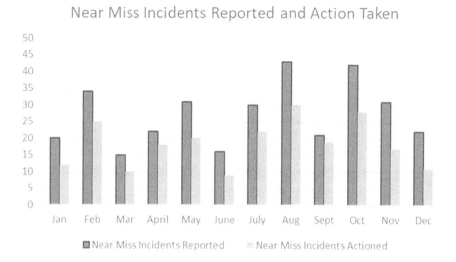

FIGURE 13.8 A graph showing the number of near miss incidents reported monthly and the number that have been actioned.

SUMMARY

The recording and investigation of near miss incidents is one of the best measures of safety management performance. Reported near miss incidents show the base of the iceberg, where the real safety problems lie. It is predictive, as near miss incidents are accident precursors. Near miss records tell an organization where they are going, while accidents tell an organization where they have been.

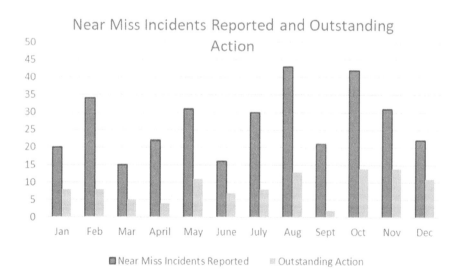

FIGURE 13.9 A graph showing the reported near miss incidents per month and the outstanding actions.

Near Miss Incidents as a Measurement of Safety Performance

Near miss incidents are accidents that the organization has not yet had. Only slightly different circumstances separate a near miss incident from a loss producing accident, so they offer ideal opportunities to fix the problem before an accident occurs.

The reporting and rectification of near miss incidents is one of the best indicators of proactive safety management action. It measures effort before consequence. By acting on near miss incidents, management heeds the warning they offer, as they are accident precursors.

To measure a near miss incident reporting and rectification system, the following criteria must be included:

- Total near miss incidents reported in the period.
- Total near miss incidents that have been risk-ranked.
- Action initiated to eliminate the near miss incident's root causes.
- Completed actions to eliminate the near miss incident's root causes.
- Feedback information on receipt and progress of near miss incident's remedial actions.

14 High-Risk Behavior (Unsafe Act) and High-Risk Conditions (Unsafe Conditions)

HAZARDS

A hazard is a situation that has potential for injury, damage to property, harm to the environment, or a combination of two or all three. A hazard is a source of potential harm. High-risk behavior and high-risk work conditions are examples of hazards.

HIGH-RISK BEHAVIOR (UNSAFE ACT)

High-risk behavior, sometimes called an unsafe act, is a departure from a normal accepted or correct work procedure, which reduces the degree of safety of that procedure.

HIGH-RISK WORKPLACE CONDITION (UNSAFE CONDITION)

A high-risk workplace condition, sometimes called an unsafe condition, is any physical or environmental condition that constitutes a hazard, and that may lead to an accident if not rectified.

IMMEDIATE ENERGY EXCHANGE CAUSES

Because of old incorrect terminology, the high-risk behaviors and high-risk conditions were termed the *immediate* causes of the accident. The reason for this carry-over is that H. W. Heinrich termed the exchange of energy segment of the accident sequence, *the accident*. This has led to confusion over the years with many still referring to an *injury* as an *accident* and an *accident* as an *injury*. High-risk behaviors and high-risk conditions are the immediate causes of the exposure, impact, or energy transfer segment of an accident.

EVENT AND CONSEQUENCE

An accident is an undesired event that results in a loss. It is a happening, an event, or occurrence. An injury, on the other hand, is a consequence of an accident and is physical harm to a person's body, health, or well-being. The high-risk behavior

High-Risk Behavior and High-Risk Conditions

FIGURE 14.1 The example accident ratio showing the high-risk behavior and conditions as the base of the triangle.

or high-risk condition, or combination, is what leads to the unplanned release and exchange of energy in the form of an exposure, impact, or energy transfer.

ACCIDENT RATIO

In reviewing the example accident ratio, high-risk behaviors and conditions form the base of the accident ratio triangle. They are the underlying causes of near miss incidents, property damage accidents, and minor and serious injuries. They lie under the waterline, and if reduced will reduce the base of the triangle and in turn reduce the number of property and injury causing accidents (Figure 14.1).

LOSS CAUSATION MODEL

The segment of the loss causation model (domino accident sequence, Figure 14.2) shows how the high-risk behavior and/or high-risk conditions lead to an inadvertent exposure, impact, or transfer of energy. Many high-risk behaviors and conditions can be identified before they result in accidental loss. These are situations that exist before the exchange of energy and, therefore, offer an opportunity to prevent injury or damage.

ACCIDENT INVESTIGATION

After an accident, it is relatively simple to identify the high-risk behaviors and conditions that lead to the transfer of energy. The accident investigation form should contain a list of these behaviors and conditions so that the investigator can check which are applicable to the accident being investigated.

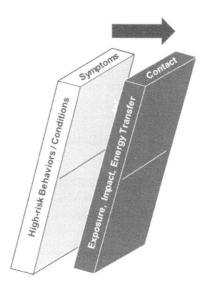

FIGURE 14.2 A segment of the loss causation model showing the domino representing the high-risk behavior and conditions that lead to the exposure, impact, or energy transfer.

High-risk behaviors and conditions are hazardous situations that exist in the workplace but not all do result in a loss, so it is beneficial to identify them before a loss occurs. The identification and rectification of these hazards is proactive safety management.

Metric

The recording and measuring of these hazards and their reduction is one method of measuring positive safety performance.

HIGH-RISK BEHAVIORS (UNSAFE ACTS)

High-risk behavior (unsafe act) is a departure from a normal accepted or correct work procedure, which reduces the degree of safety of that procedure.

These behaviors can be described as any activity carried out by workers which is not according to the prescribed safety standard, or practice, and which can cause accidents, or have the potential to cause accidents. Such high-risk behaviors may be due to poor attitude of workers, lack of awareness of safety measures, or not following job safe practices (JSPs) as a result of other hidden or root causes. High-risk behaviors may be as a result of human failure, errors, or violations.

A Safety Myth

For many years, unsafe acts or high-risk behaviors of workers have been cited as the cause of the majority of work accidents. Some philosophies state that up to 90% of

High-Risk Behavior and High-Risk Conditions

accidents are caused by high-risk worker behavior. Many of these statements cannot be supported by research, and are a carryover of H. W. Heinrich's research, which stated that 88% of accidents were caused by the unsafe acts of people. This is not a substantiated fact. This research was done in 1929 and has never been substantiated but is still believed today.

High-risk behaviors are one of the immediate causes of the unplanned impact and transfer of energy that results in a loss. (The other category of immediate causes is the high-risk workplace conditions.) The following are the most common categories of high-risk behavior:

- Operating equipment without authority or training.
- Failure to warn.
- Failure to secure.
- Improper lifting.
- Not following procedures.
- Working at unsafe speed.
- Removing safety devices or making safety devices inoperative.
- Using unsafe or defective equipment.
- Using equipment improperly.
- Unsafe placing.
- Taking up an unsafe position or unsafe positioning.
- Working on or servicing moving or dangerous equipment.
- Distracting, teasing, or horseplay.
- Failure to wear personal protective equipment (PPE).
- Improper or unsafe positioning.
- Improper loading.

HIGH-RISK BEHAVIOR IS NOT A NEAR MISS INCIDENT

Numerous high-risk behaviors are committed daily but do not result in a contact with energy, or impact of any sort. These are not near miss incidents, as there has been no flow of energy that, in a contact situation, would have caused injury, damage, or other loss. Many confuse the high-risk behavior and the high-risk workplace condition with near miss incidents. To fall into the latter category, there must be a flow of energy and potential for an exchange of energy.

A COMPLEX SITUATION

The term "high-risk behavior" or "unsafe act" describes a complex situation. High-risk behavior is not merely a worker blatantly defying safety rules and regulations. It is the end action that results from an accumulation of a number of breakdowns and weaknesses in a management system, which was designed to keep the worker safe at work.

Organizations should look beyond the high-risk actions uncovered by accident investigation and seek the deep-rooted causes for these actions.

HUMAN FAILURE

Human failure is a broad term often used when an accident occurs. There are two main types of human failure, *inadvertent failure* (error) and *deliberate failure* (violation).

INADVERTENT FAILURE

Inadvertent failures are classified as mistakes that could be rule-based or knowledge-based mistakes. Workers do make mistakes, but they are not the prime accident cause.

DELIBERATE FAILURE

Deliberate failures are violations. These violations could be routine or could be as a result of a certain situation or exceptional circumstance.

EXCEPTIONAL FAILURE

Exceptional failure is where a person attempts to solve a problem in highly unusual circumstances and takes a calculated risk by breaking the rules.

ACTIVE AND LATENT FAILURES

High-risk behaviors are sometimes referred to as *active failures*. These are errors and violations that have an immediate negative result. Active failures have an immediate consequence and are usually made by frontline people such as drivers, control room staff, or machine operators. In a situation where there is no room for error, these active failures have an immediate impact on health and safety.

Latent failures are made by people whose tasks are removed in time and space from operational activities, for example, designers, decision makers, and managers. Latent failures are typically failures in health and safety management systems (SMSs) (design, implementation, or monitoring). Examples of latent failures are as follows:

- Poor design of plant and equipment.
- Ineffective training.
- Inadequate supervision.
- Ineffective communications.
- Uncertainties in roles and responsibilities.

ERRORS

An error is an action that fails to produce the expected result, and that may produce an undesired and unwanted outcome. Human error is commonly defined as a failure of planned action to achieve a desired result. Some categories of error are as follows.

Slips or Lapses

Slips or lapses are unplanned actions. They are unintended actions that sometimes occur when the wrong step is taken, or due to a lapse, a step of a procedure or process is not done correctly or is left out.

A *slip* happens when a person is carrying out familiar tasks automatically, without thinking, and the person's action is not as planned, such as operating the wrong switch on a control panel.

A *lapse* happens when an action is performed out of sequence or a step in a sequence is missed.

Mistakes

Mistakes are made when decisions and actions are taken and later discovered to be incorrect, although the employee, at the time, thought they were correct. They are failures in a plan of action. Even if the execution of the plan was correct, it would be impossible to achieve the desired outcome.

Rule-based mistakes happen when a person has a set of rules about what to do in certain situations and applies the wrong rule.

Knowledge-based mistakes happen when a person is faced with an unfamiliar situation for which he or she has no rules, uses his or her knowledge, and works from first principles, but comes to a wrong conclusion.

Latent Errors

Latent errors are problems, or traps hidden within systems, which under certain conditions will contribute to an error occurring. They may lie dormant for some time but given a certain set of circumstances they manifest.

Violations

Violations are deliberate deviations from safe work standards and procedures. They can be accidental, unintentional, or deliberate. Violations are rule-breaking actions and are deliberate failure to follow the rules, such as cutting corners to save time or effort, based on the belief that the rules are too restrictive and are not enforced anyway.

Routine Violations

Routine violations are identified in most violation categories. Routine violations occur when the normal way of doing the work is different from prescribed rules and procedures. Often routine violations are so common amongst work teams, that they are no longer perceived as violations or high-risk behaviors. This is called "that's the way things are done around here."

Unintentional Violations

Unintentional violations occur when rules are written that are almost impossible to follow. This could occur when workers do not know or understand the rules that they are expected to follow.

Procedural Violations

Procedural violations happen when procedures are purposefully deviated from ignored or bypassed. This is often summarized as "failure to follow procedures." The reason for this may be that the procedure is incorrect, outdated, or difficult to follow.

Exceptional Violations

Exceptional violations occur when an isolated departure from procedure occurs. This type of violation is neither typical of the employee, nor condoned by supervision. Exceptional violations occur in unusual circumstances. In some crisis situations, these violations may even be inevitable, especially when it is believed that the violation is necessary to cope with the exceptional circumstances.

Situational Violations

Situational violations occur when circumstances in the workplace, such as time, pressure, or a sense of urgency require, or encourage employees to violate safety rules.

HIGH-RISK WORKPLACE CONDITIONS

A high-risk workplace condition (unsafe condition) is any physical condition that constitutes a hazard, and that has the potential to lead to an accident if not rectified. Any condition or situation (electrical, chemical, biological, physical, mechanical, and environmental) that increases the risk and possibility of accidents can be called a high-risk condition.

Although there are numerous high-risk workplace conditions, the following are common categories of high-risk conditions:

- Unguarded, absence of required guards.
- Inadequately guarded.
- Defective, rough, sharp, or cracked.
- Unsafely designed machines or tools.
- Unsafely arranged and poor housekeeping.
- Congested/restricted/overcrowded work area.
- Inadequate or excess illumination/sources of glare.
- Inadequate ventilation.
- High or low temperature exposure.
- Unsafely clothed, no PPE, inadequate or improper PPE.
- Unsafe process.
- Noise exposure.
- Radiation exposure.
- Fire and explosion hazard.
- Inadequate warning system.
- Hazardous environment.

IDENTIFYING AND MEASURING HAZARDOUS SITUATIONS

Inspections, walk abouts, reviews, and observations are methods of hazard identification. Incident recall sessions can also be used to recall hazardous situations that may not have been detected by other methods. The main indicator is the number of hazards identified and the number of hazards that have been reduced or eliminated.

SAFETY OBSERVATION PROGRAM

A safety observation program is one program within a complete SMS and includes the following elements.

Appointed Observer(s)

Designate a person to conduct the safety observation. This may be a supervisor or employee who has undergone training on how to complete an observation effectively. The safety personnel can also conduct observations.

Checklist

Safety observations should include high-risk behaviors as well as high-risk workplace conditions and usually follow a checklist organized by subject area or combination such as:

- High-risk behaviors:
  - Wearing of PPE.
  - Following procedures.
  - Correct positioning, etc.
- High-risk conditions:
  - Housekeeping.
  - Machine guarding.
  - Machine usage.
  - Work at height, etc.

Time of Observation

A scheduled time must be allocated to the observation tour. Depending on the size of the area being checked, time must be allocated to complete the safety observation.

Observations Process

The observers conducting the safety observation observe employees while they are performing their normal daily tasks. The timing of the observation is important and should form part of the observer's training on how to conduct observations.

SAFETY OBSERVATION CARD

To guide the observer a checklist in the form of an observation card should be provided and used for all observations. This acts as a guide as to what to observe and provides for immediate recording of the observation. The card (Figure 14.3) includes provision to record the time and date of the observation, the number of safe observations, and the number of unsafe observations.

OBSERVATIONS

Observations should not be made only of employees' actions but also of the workplace safety conditions. It would be futile to record an employee not wearing eye protection when the housekeeping in the area is hazardous and the fire extinguishers are blocked and inaccessible. All hazards should be recorded as they are noticed.

THE OBSERVATION PROCESS

A trained, controlled group of observers make observations of employees at work during normal shifts. These observations note the date and time, location, and number of employees observed. Each observation will be designated safe or high risk, and if high risk, observers should be trained to constructively correct the high-risk behaviors and provide positive feedback on safe behaviors. Safety observations are based on people or conditions observed and not behaviors observed. A maximum of six to seven employees should be observed during one session.

Safety Observation Card	
High-risk☐	Safe Work☐
Observer:	
Date:	
Time:	
Area/Section:	
Supervisor:	
Number of Observations: 1 2 3 4 5	
☐Eye PPE	☐Stacking
☐Foot PPE	☐Energy Control
☐Fall PPE	☐Unsafe Speed
☐Height PPE	☐Lifting
☐Confined Space	☐Positioning
☐Trip Hazards	☐Housekeeping
☐Tool Safety	☐Permits
☐Walkways Clear	☐Safe Work

FIGURE 14.3 An example of a safety observation card.

High-Risk Behavior and High-Risk Conditions

Discussions

Safety observations include conversations with employees to inform them of safety issues noted, to ask questions, or to commend them for following the correct safety procedures.

Record Findings

The findings noted during the observation tour should be recorded and these should list good conditions and behavior noted as well as deviations from accepted practices.

Tracking System

On a weekly basis, the observations are input into a tracking database from which the trends are determined. The results of the weeks' observations are communicated both to employees and management via the normal communication methods of newsletters, safety noticeboards, and electronic feedback.

Positive as well as negative observations from the previous weeks' data are identified and emphasized to management and the workforce. Recognition should be given for safe behavior observed.

Observation Metrics

$$\% \text{ Safe observations} = \frac{\text{Safe observations}}{\text{Total observations}} \times 100$$

$$\% \text{ High-risk observations} = \frac{\text{High-risk observations}}{\text{Total observations}} \times 100$$

$$\% \text{ Corrective actions taken} = \frac{\text{High-risk observations rectified}}{\text{Total observations}} \times 100$$

Report

At the end of the period (week or month), a summary of the observation findings should be summarized and circulated to all. This figure is a proactive leading safety performance metric and should be discussed at safety committee meetings. A summary sheet (Figure 14.4) can be used to summarize the results for the period.

The report could also be presented in a graphic form for circulation and discussion as in the example (Figure 14.5).

INCIDENT RECALL

The objective of an accident and near miss incident recall program is recalling unreported events, high-risk behaviors, high-risk conditions, and near miss incidents that did not result in any visible injury, damage, or production loss, but which may have if

Safety Observation Report for Month:	%	Number
Total Observations	100%	485
% High-risk Behavior	33%	160
% High-risk Conditions	23%	111
% Safe Conditions	44%	213
Most recurring High-risk Behavior:		
Most recurring High-risk Condition:		

FIGURE 14.4 A weekly or monthly observation summary sheet.

circumstances were different. The objective is to learn from these events. It can also be used to review and discuss past accidents.

GOAL OF INCIDENT RECALL

A further goal is to systematically gather information and learn from near miss incidents and accidents that may not have been reported, so that the information can be shared, and future events prevented. Incident recall is the ideal way of reminding people. By recalling past injury or damage causing accidents, high-risk conditions, and behaviors, and bringing about an awareness of their causes will help to ensure that steps are in place so that a recurrence does not happen.

SUMMARY

A hazard is a situation that has potential for injury, damage to property, harm to the environment, or a combination of two or all three. A hazard is a source of potential harm. High-risk behavior and high-risk work conditions are examples of hazards.

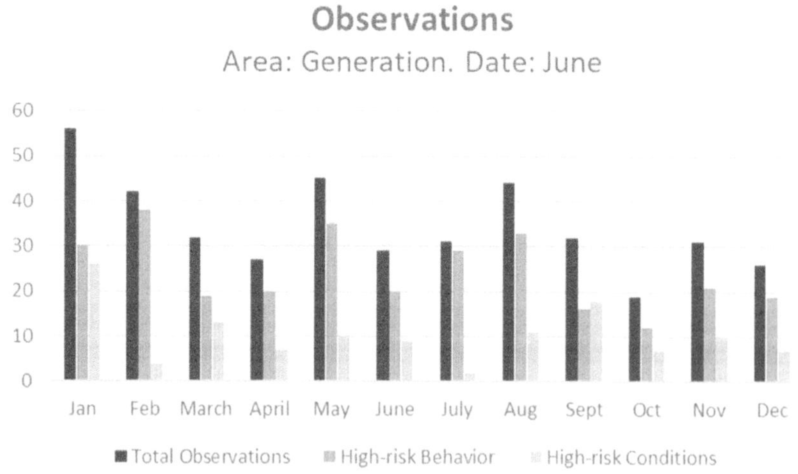

FIGURE 14.5 A graph showing total monthly observations, high-risk behavior, and high-risk workplace conditions recorded.

High risk behavior, sometimes called an unsafe act, is a departure from a normal accepted or correct work procedure, which reduces the degree of safety of that procedure. A high-risk workplace condition, sometimes called an unsafe condition, is any physical or environmental condition that constitutes a hazard, and that may lead to an accident if not rectified.

An accident is an undesired event that results in a loss. It is a happening, an event, or occurrence. An injury, on the other hand, is a consequence of an accident and is physical harm to a person's body, health, or well-being. The high-risk behavior or high-risk condition, or combination, is what leads to the unplanned release and exchange of energy in the form of an exposure, impact, or energy transfer.

An inspection and observation system, as well as an incident recall program, should form part of the SMS and the findings should be tabulated and reported to management at regular intervals. The number of hazards recorded and the number of remedial measures taken are vital leading indicators of safety management.

15 Health and Safety Inspections

KEY PERFORMANCE INDICATORS

Safety management performance measurement is an essential management tool, as it helps us determine if the organization's safety effort is making an impact, correctly managing resources, and helps focus improvement efforts.

Key performance indicators (KPIs) are the critical (key) quantifiable indicators of progress toward the intended result of a healthy and safe workplace. KPIs provide a focus for strategic and operational improvement. They create an analytical basis for decision-making and help focus attention on what matters most. Managing with the use of KPIs includes setting targets (the desired level of performance) and tracking progress against those targets.

Managing with KPIs often means working to improve performance using leading indicators, which are precursors of future success, that will later drive desired impacts indicated by lagging measures.

Each element of the safety management system (SMS) should be viewed as a KPI as they are targets for achievement, can be managed, measured, and quantified. The process of setting health and safety standards and measuring against those standards can be applied to all elements of the SMS and forms the basics of safety auditing.

INSPECTIONS

Inspections are an organization's first line of defense in proactively preventing workplace accidents. Safety inspections are proactive safety management actions and are a good leading indicators of safety management performance. Both the number and the results of inspections are used as indicators of positive performance.

To be a meaningful performance metric, the quantification safety inspections should include following:

- Total number of inspections.
- Number of hazards identified.
- Number of hazards corrected.

A safety inspection is *a monitoring function to locate, identify, and eradicate high-risk conditions (hazards) and high-risk behaviors (hazards), which have the capacity to lead to accidental losses.*

Health and Safety Inspections

PURPOSE OF SAFETY INSPECTIONS

Safety inspections are important as they allow the inspector to:

- Identify existing and future hazards.
- Determine underlying causes of hazards.
- Listen to the concerns of workers and supervisors.
- Gain further understanding of jobs and tasks.
- Recommend corrective action.
- Monitor the steps taken to eliminate hazards.
- Meet regulatory and SMS requirements.

An inspection involves a tour around the physical work environment with the specific objective of ensuring the health and safety of the workers, products, equipment, and machinery in that area.

The inspection not only includes the work area itself but also the process being carried out, the movement of material, product and finished goods, as well as the actions, working conditions, and general safety of the employees. Short cuts, unsafe situations, and high-risk behaviors are also detected during an inspection. Once these have been identified, positive remedial measures can be taken to rectify them as part of the proactive safety process.

TYPES OF HAZARDS

Hazards can occur due to high-risk workplace conditions and practices in the workplace. Types of workplace hazards that can be identified during safety inspections include following:

Safety Hazards

Safety hazards are caused by:

- Inadequate machine guards.
- Hazardous energy (mechanical, electrical, gravitational, pneumatic, etc.).
- Vehicles.
- Machinery.
- Tools.
- Lack of fall protection.
- Confined spaces.
- Housekeeping, etc.

Biological Hazards

Biological hazards are caused by organisms such as insects, viruses, bacteria, fungi, and parasites.

Chemical Hazards

Chemical hazards are caused by a solid, liquid, vapor, gas, dust, fume, or mist.

Ergonomic Hazards

Ergonomic hazards are caused by poorly designed workstations, tools and equipment, improper work methods, and incorrect manual material handling. These place physiological (repetitive and forceful movements, awkward postures, or overloading) and psychological (workload or time pressure) demands on the worker that can lead to musculoskeletal injuries.

Physical Hazards

Physical hazards are caused by heat, cold, noise, vibration, radiation, pressure, odors, and indoor air quality.

Psychosocial Hazards

Psychosocial hazards are hazards that can affect the mental health or well-being of employees and include overwork, stress, bullying, or violence and harassment.

HAZARD IDENTIFICATION

The range of activities undertaken by an organization will create hazards, which will vary in nature and significance. The range, nature, distribution, and significance of the hazards are called the *hazard burden* and will determine the risks that need to be controlled.

Hazard Classification

A simple hazard classification system is the A, B, and C classification:

- A – Likely to cause death, permanent disability, extensive property damage, or even catastrophic results.
- B – Likely to cause serious injury but less serious than an A class hazard, substantial property loss, or damage to the environment.
- C – Likely to cause minor injury, relative property damage, and minor disruption.

Hazard Profiling

Hazard profiling is a process of describing the hazard in its local context, which includes a general description of the hazard, a local historical background of the hazard, local vulnerability, possible consequences, and estimated likelihood, and is very similar to risk assessment.

Health and Safety Inspections 159

MEASURING AGAINST STANDARDS

What gets measured, gets done, consequently, a safety inspection is the ideal method of measuring the standards in the workplace against the SMS and legal standards. An effective method of measuring compliance to standards is by conducting a formal, scheduled safety inspection.

Unplanned inspections are sometimes referred to as management by walking around (walkabout), and is a style of business management which involves managers wandering around, in an unstructured manner, through the workplace, to check on employees, equipment, or on the status of health and safety in the workplace.

CHECKLIST

After each inspection, there should be a written report or a completed checklist. This should be completed as soon as possible after the inspection and submitted to the relevant departments or supervisors for deficiencies to be rectified. The checklist is a vital document in the inspection process and some form of checklist or guideline should be used for all inspections. A safety inspection checklist structures the inspection and makes it meaningful and more thorough.

INSPECTION GUIDELINES

There are certain guidelines for carrying out a safety inspection and supervisors and employees have been de-motivated because of having had their work or production area inspected by inspectors using the wrong approach. The wrong approach, a negative attitude, or incorrect use of certain selected words can be de-motivating.

Guidelines for conducting a successful inspection are as follows:

- Always use a positive approach during the inspections and emphasise the fact that the inspection mission is a fact-finding and not a fault-finding exercise.
- Always greet the person whose area is being inspected in a friendly manner and thank them for taking the time to accompany the inspector.
- Commend the person on the positive aspects found during the inspection and ask their opinion of the deviations that may have been noted.
- Endeavour to have major hazards rectified immediately, as a hazard should never be left without some form of rectification having taken place.
- The inspection should be a pleasurable experience by the person whose area is inspected. They should look forward to gaining further insight and assistance during the next inspection.
- The attitude and diplomacy of the person doing the inspection is of prime importance.
- During inspections, some organizations hand out tokens of appreciation to individuals or supervisors who have made an exceptional effort in maintaining a safe, neat, and healthy work area.
- Always thank the person at the end of the inspection.

TYPES OF INSPECTIONS

HEALTH AND SAFETY REPRESENTATIVES' INSPECTIONS

Health and safety representatives are appointed because of their knowledge and familiarity with the work environment and work process. No one is better equipped to carry out a safety inspection of a particular area than the person who works in the area, knows the process, and has many years of experience in their work environment. Health and safety representatives also prove invaluable as their inspection techniques are very thorough and no quarter is overlooked during their inspections.

Although they may not have the authority to take remedial action on certain hazards noted, the health and safety representatives' reports must receive prompt attention by management to prevent the same hazards from being reported repeatedly. Once it appears that no actions are forthcoming from the safety representatives' inspection reports, they could become disillusioned, lose heart and the benefits of their inspection system could be lost.

Involvement, and inspections, of areas by health and safety representatives is based on the principle of involvement in safety that subsequently motivates the people who are involved. Constant reinforcement, support, and commendations should be given to the safety and health representatives concerning their inspections, and an ongoing system of encouragement should be in place (Figure 15.1).

RISK ASSESSMENT INSPECTION

A risk assessment of hazards cannot be carried out successfully without first conducting an on-site inspection of the physical conditions, the raw products used, the processes, and machinery and transportation modes. Only once the entire process and exposures have been physically examined by an inspection, can a risk assessment of the hazards be compiled. A risk assessment inspection will assist the process of identifying hazards and assessing their risks.

LEGAL COMPLIANCE

A legal compliance inspection involves an inspection to confirm compliance to regulations and laws governing the physical work conditions. Safety laws and regulations also require certain regular inspections of workplaces and equipment.

INSPECTION TYPE	FREQUENCY	MONTHLY TARGET	ACTUAL	PERCENTAGE	DEVIATION
Health and safety representative	Monthly	35	30	85%	−15%
Area inspection	Weekly	8	8	100%	0%
Safety department	Weekly	10	9	90%	−10%
Management	Monthly	1	1	100%	0%
Housekeeping	Monthly	4	2	50%	−50%

FIGURE 15.1 A table showing the type of inspection, the monthly target, the actual achievement, and deviation from target.

Health and Safety Inspections

Informal Walk About

An informal inspection should take place on a regular basis, preferably daily. This would be a walk about by the supervisor of that work area and should cover all the area under his or her jurisdiction. The informal inspection can identify hazards, which can be immediately rectified.

Planned Inspections

Planned inspections are inspections that are planned on a regular basis. They normally follow a predetermined route and are carried out by a team comprising safety personnel, the supervisor of the area, and the health and safety representative of that area.

These inspections are normally scheduled, a certain date is allocated (weekly, monthly, or three-monthly) and the areas nominated for inspection are determined in advance. Planned inspections are an ideal way of covering the entire work area, as they are a systematic approach to covering all areas of the workplace.

Safety Observation

Safety observations inspections should include high-risk behaviors as well as high-risk workplace conditions and usually follow a checklist organized by subject area. These could be formal or informal inspections.

Safety Department Inspection

Being thoroughly familiar with the work environment, a regular inspection carried out by the members of the safety department is of vital importance. It is also a systematic approach to ensuring that all hazards, deviations from laid-down standards, and also high-risk behavior and conditions are identified and rectified on an on-going basis.

Housekeeping Competition Inspection

A separate inspection may be conducted to adjudicate the internal housekeeping competition. This process is facilitated by a physical inspection of the different work areas.

Business order (good housekeeping) is the foundation of an SMS and should be inspected on a regular basis. The example checklist (Figure 15.2) shows an extract of a housekeeping checklist which lists the details of the standard and the (0–5) scoring method for each detail.

Safety Survey

A safety survey is a thorough safety inspection of the work environment and includes an inspection of control systems in place. A safety survey will highlight the most prominent deficiencies in the physical conditions and the safety management control systems.

HOUSEKEEPING INSPECTION SCORE SHEET						
Floors						
Are the floors safe and free from slip or trip (cords) hazards?	0	1	2	3	4	5
Are they clean, dry, correctly marked, and clearly demarcated?	0	1	2	3	4	5
Are scraps and refuse bins provided for removal of superfluous material?	0	1	2	3	4	5
Yard, outlying areas						
Are they in good order, free of excessive material, and junk?	0	1	2	3	4	5
Demarcation						
Are no parking and no stacking areas demarcated with a standardized color code?	0	1	2	3	4	5
Are the demarcations being adhered to?	0	1	2	3	4	5
Is the demarcation clearly visible?	0	1	2	3	4	5
If the demarcation has been carried out by other means, are the areas clearly indicated to eliminate confusion?	0	1	2	3	4	5

FIGURE 15.2 An example of a housekeeping inspection checklist and score sheet (abbreviated).

Safety Audit Inspection

A safety audit inspection is a thorough inspection of the physical workplace. It is followed by a documentation review wherein the physical work conditions noted are compared with the SMS standards and quantified. An audit inspection is a part of an audit but is not an audit on its own.

Specific Equipment Inspections

Specific equipment inspections are routine inspections of items of equipment that are used regularly, and that are essential to ensure the safety and well-being of employees. Special equipment inspections are scheduled periodic inspections of the equipment determined by the frequency of use and consequences of equipment failure. They can include following:

- Firefighting equipment.
- Motorized transport.
- Fixed electrical installation.
- New equipment.
- Ladders.
- Pressure vessels.
- Boilers.
- Lifting gear, etc.

Other Inspections

Safety inspections can vary greatly from workplace to workplace and could include following:

- Ergonomic inspections.
- Accident near miss incident investigation inspections.

Health and Safety Inspections

- Legal inspections by the local legal entities.
- Occupational hygiene inspections.
- Housekeeping inspections.
- Safety competition inspections.
- Contractor's site inspection.
- Self-audit inspections.
- Risk management inspections by insurers, etc.

CONCLUSION

Correctly planned and executed health and safety inspections should be carried out on a regular basis by designated employees to identify workplace hazards and to have them rectified before an accidental loss occurs. In measuring the efficiency of these inspections, the number of inspections should be measured as well as the corrective actions taken as a result of the hazards identified during the inspection.

SUMMARY

Inspections are an organization's first line of defense in proactively preventing workplace accidents. Safety inspections are proactive safety management actions and are a good leading indicator of safety management performance. Both the number and the results of inspections are used as indicators of positive performance.

The range of activities undertaken by an organization will create hazards, which will vary in nature and significance. The range, nature, distribution, and significance of the hazards are called the *hazard burden* and will determine the risks that need to be controlled.

What gets measured, gets done, consequently, a safety inspection is the ideal method of measuring the standards in the workplace against the SMS and legal standards. An effective method of measuring compliance to standards is by conducting a formal and scheduled safety inspection.

16 Hazard Identification and Risk Assessment

The prime function of safety management is to identify the hazards in the workplace, assess them for risk of occurrence and severity, and apply the hierarchy of control to reduce or eliminate those risks. This safety management activity can be measured as an important leading indicator of safety performance.

THE PURPOSE OF HAZARD IDENTIFICATION AND RISK ASSESSMENT

The purpose of hazard identification and risk assessment (HIRA) is to identify hazards and evaluate the risk of injury or illness arising from exposure to these hazards with the goal of eliminating the risk, or using control measures to reduce the risk at the workplace. An HIRA process includes the following steps:

- Identify the hazards.
- Do a hazard classification exercise.
- Conduct a risk analysis considering probability, severity, and frequency.
- Allocate risk scores.
- Rank the risks according to the scores.
- Evaluate the risks.
- Compile a risk profile.
- Identify controls to be put in place, with consideration given to how effective and adequate the proposed controls would be.
- Review the assessment and update if necessary.

SOURCES OF HAZARDS AND HAZARD BURDEN

Sources of hazards should be identified and documented in a risk register that is modified and updated regularly. This will direct the design and requirements of the health and safety management system (SMS) as the system must be focused around reducing risks arising from the workplace and its activities.

MEASURING THE HAZARD BURDEN

Identifying hazards and the hazard burden within the organization is the first step in the HIRA process. A number of hazard identification methods can be used, and as with other safety processes, these are ongoing activities.

When measuring the hazard burden, the following questions should be asked:

- What are the hazards associated with the company activities?
- What is the significance of the hazards (high/low)?
- How does the nature and significance of the hazards vary across the different parts of the organization?
- How does the nature and significance of the hazards vary with time?
- Is the organization succeeding in eliminating or reducing hazards?
- What impact are changes in the business having on the nature and significance of hazards?

Ongoing Assessments

The organization should perform ongoing and documented HIRAs for the work at hand. The assessments should identify competencies required of the employees, as well as controls and barriers required to guard against identified hazards. The assessments should include identifying which company policies, standards, procedures, and processes apply to the workplace situations. The organization's health and safety department may be consulted to establish a risk ranking of the various work processes.

HIERARCHY OF CONTROL

There should be multiple layers of controls protecting employees from hazards and associated risks. The types of controls could include the following:

- Elimination – which would mean changing the way the work is to be done.
- Engineering controls – such as the re-design of a worksite, equipment modification, and tool modification.
- Administrative controls – such as a change of work methods and re-scheduling of work.
- Health and safety system controls – such as health and safety policies, SMS standards, work procedures, training, and personal protective equipment (PPE) (SMS elements).

RISK ASSESSMENT

To help determine the potential probability, frequency, and severity of loss which could be caused by hazards, the technique of risk assessment is a major element in the health and SMS. The purpose of HIRA is as follows:

- To identify all the pure risks within the organization and which are connected to the operation (*hazard identification*).
- To do a thorough analysis of the risks taking into consideration the frequency, probability, and severity of consequences (*risk analysis*).

- To implement the best techniques for risk reduction (*risk evaluation*).
- To deal with the risk where possible (*risk control*).
- To monitor and re-evaluate on an ongoing basis (*re-assessment*).

RISK REGISTER

A risk register (Figure 16.1) is a documented risk tracking process. It is a management tool that records risk, records the possible risk outcomes, prescribes control methods, and allocates responsibility for risk reduction actions.

The risk register should contain:

- Risk ID – a unique tracking number.
- Risk category – the type of risk.
- Risk description – a brief description of the risk.
- Possible likelihood and severity – a risk ranking based on a risk matrix.
- Risk owners – who is responsible for managing the risk?
- Risk scores – what is the risk ranking?
- Mitigation actions – actions required.
- Completion date – date of expected completion of actions.
- Completed – date on which risk declared as low as is reasonably practicable (ALARP).

APPLICATION OF THE HIRA PROCESSES

HIRA should be ongoing processes within the SMS of the organization to ensure correct control measures are implemented to reduce the risk created by workplace hazards. Since these are leading activities within the SMS, they can be measured.

These measurements could include the following:

- The number of Hazard and Operability Studies (HAZOPS) conducted.
- Ongoing critical task identification processes.
- An active change management program.
- Ongoing diverse health and safety inspections.
- Effective near miss and accident investigations.
- Daily task risk assessments conducted.

AUDIT OF AN HIRA PROCESS

The audit inspection will indicate the effectiveness of the HIRA process. If a number of hazards are noted, this will indicate that the system is not working satisfactorily. The fewer hazards found, the better the system is functioning. Hazard identification should be incorporated into all inspections, and an ongoing observation reporting system should be in place, so that reported hazards can be identified and eliminated. Incident recall sessions should include information about hazards. Safety training in hazard recognition should take place frequently.

Hazard Identification and Risk Assessment

#	Risk Description	Category	Probability	Severity	Score	Action Plan	Date	Owner	Date Completed
121	Uneven floor in walkway at storage.	HK	M	L	4	Floor tiles to be removed and re-laid.	07/12/23	Jim B.	
122	Tail end pulleys at coal conveyor unguarded.	MG	MH	H	12	Install correct machine guard.	04/01/23	Davis G.	
123	Illumination in maintenance shop below accepted levels.	ILL	L	L	1	Illumination survey to be conducted and maintenance to install or repair lighting up to required standard.	04/01/23	IH and Elect Maintenance.	
124	Arc flash labeling not completed.	AF	MH	S	15	Label survey to be completed in full by end Feb. 2003.	02/28/23	Elec. Maintenance.	
125	Procedure for trench shoring not completed.	MG	M	H	8	Procedure to be written and approved.	05/09/23	Civil Department.	
126	Noise zoning survey incomplete.	ENV	L	MH	4	Complete zoning and implement HC program in full.	06/30/23	IH Department.	

FIGURE 16.1 An example of a risk register.

AUDIT PROTOCOL

The most effective method of measuring the SMS process of HIRA is to audit the performance against the company SMS standard for that process. A proposed HIRA audit protocol (Figure 16.2) contains the following:

- The SMS element, program, or process description.
- The maximum points allocated to each minimum standard requirement.
- Questions that could be asked during the audit.
- Verification documentation required.
- What to look for during the on-site inspection?

Each criterion is scored on a (0–5) scoring system where:

1 = 20%
2 = 40%
3 = 60%
4 = 80%
5 = 100%

KEY SAFETY PERFORMANCE INDICATOR

Due to the criticality of this safety process, HIRA should be a key performance indicator (KPI) for managers at different levels within the organization. By identifying hazards, assessing the risks, and applying risk reduction controls, the company will be taking positive actions to prevent workplace accidents. This ongoing process can be monitored and measured as a leading indicator of safety management performance.

SUMMARY

The purpose of HIRA is to identify hazards and evaluate the risk of injury or illness arising from exposure to a hazard with the goal of eliminating the risk, or using control measures to reduce the risk at the workplace.

Identifying hazards and the hazard burden within the organization is the first step in the HIRA process. A number of hazard identification methods can be used, and as with other safety processes, these are ongoing activities.

The organization should perform ongoing documented HIRAs for the work at hand. The assessments should identify competencies required of the employees, as well as controls and barriers required to guard against identified hazards. To help determine the potential probability, frequency, and severity of loss that could be caused by hazards, the technique of risk assessment is a major element in the health and SMS.

HIRA should be ongoing processes within the SMS of the organization to ensure correct control measures are implemented to reduce the risk created by workplace hazards. Since these are the leading activities within the SMS, they can be measured.

The most effective method of measuring the SMS process of HIRA is to audit the performance against the company SMS standard for that process.

Hazard Identification and Risk Assessment

ELEMENT / PROGRAM / PROCESS	POINTS	SCORE	QUESTIONS THAT COULD BE ASKED	VERIFICATION	WHAT TO LOOK FOR
HAZARD IDENTIFICATION AND RISK ASSESSMENT (HIRA)					
HAZARD IDENTIFICATION					
Hazards are identified	5		What hazard identification methods are used?	See examples	Cross check with hazards noted during inspection
Inspections	5		How often are inspections done?	Copy of a completed checklist	Is the entire area covered?
Walkabouts	5		How often are walkabouts done?	Copy of a completed checklist	Proof of hazard identification
Checklists	5		Are checklists used?	Copy of a completed checklist	Is it applicable?
Incident recall sessions	5		How often are inspections done?	Copy of minutes/records	Are hazards being recalled at these sessions?
RISK ASSESSMENT					
Risks assessed? Daily task RA's	5		How are risks assessed?	Example	Check for probability/severity and ranking
Matrix used	5		Is a risk matrix used?	Example of completed matrix	Is it being used?
Risk reduction efforts	5		What risks have been reduced during the last month?	Examples of risk abatements	Is the system working?
Risk register	5		Is a risk register kept?	Inspect register. Up to date?	Is the register kept evergreen?
TOTAL	45				

FIGURE 16.2 The audit protocol for the element hazard identification and risk assessment. (From McKinnon, R. C. 2020. *The Design, Implementation and Audit of Safety Management Systems* (Figure 13.2). Boca Raton: Taylor and Francis. With permission.)

17 Health and Safety Committee Meetings

Health and safety committees are formally constituted committees consisting of a cross representation of management, employees, and worker representation. They have a constitution that describes their functions and duties and guides their activities. A meeting of employees and safety departments to discuss safety issues is not a formal health and safety committee as prescribed by the health and safety management system (SMS) requirements.

MEASURABLE MANAGEMENT PERFORMANCE

The constituting of and functioning of a safety committee is a measurable management performance and managers at all levels should be responsible for establishing and maintaining committees within their areas of responsibility. The process of establishing and maintaining committees can be measured on a monthly basis and be regarded as an important safety key performance indicator (KPI).

A HEALTH AND SAFETY COMMITTEE

A health and safety committee can be defined as *a group consisting of management, employees, and worker representation that aids and advises the organization on matters of health and safety pertaining to company operations.* It is a forum for joint consultation on safety and it performs essential monitoring, educational, investigative, and evaluation tasks within the framework of the SMS.

JOINT HEALTH AND SAFETY COMMITTEE

A joint health and safety committee is advocated by health and safety legislation in most countries. The joint committee concept stresses co-operation and a commitment to safety as a shared responsibility by both management and employees. The concept of joint decision-making is applied as management and workers can now face each other across the same meeting table with a common ground and a common agenda based on the prevention of accidents.

HEALTH AND SAFETY REPRESENTATIVES' COMMITTEE

In some countries, a separate committee consisting of managers and health and safety representatives is required by health and safety regulations. During these meetings, the inspection reports of the representatives are tabled and discussed. Management then draws up action plans to rectify the deviations found by the inspections and

Health and Safety Committee Meetings

reports back at subsequent meeting as to what actions have been taken to eliminate the hazards reported. This is an ongoing process.

FUNCTIONS OF HEALTH AND SAFETY COMMITTEES

The functions of a health and safety committee are as follows:

- To meet regularly.
- To ensure that health and safety momentum is maintained.
- To provide two-way communication.
- To solve certain health and safety problems.
- To monitor the progress of the SMS.
- To assist in the review and approval of SMS standards and procedures.
- To function as a selection committee for suppliers of personal protective equipment (PPE).
- To assist in accident or near miss incident investigations.
- To sort and select safety suggestions.
- To assist in planning safety campaigns, competitions, etc.
- To discuss injury statistics and monitor trends.
- To review internal and external audit results.

LEADERSHIP

To be effective, the committee meeting must be chaired by the manager of the department or division where the committee is constituted. Often the mistake is made of having the safety department head up the committee meeting which detracts from the purpose and effectiveness of the committee. Since the committee is established to provide safety communication between management and other levels within the organization, managers must chair the meetings.

THE COMMITTEE SYSTEM

Health and safety committees should be established on a tier system. Within an organization, there should be an executive safety committee at the executive management level, and representatives of this committee then hold similar meetings at committees established within their divisions. Attendees of these divisional committees then, in turn, hold committee meetings at lower levels within their departments. This will create a cascading effect where information, decisions, and reports are cascaded down to committees at lower levels ensuring communication throughout the organization (Figure 17.1).

MEETING AGENDA

A typical agenda for a health and safety committee meeting would contain the following agenda items:

- Notice of meeting.
- Apologies.

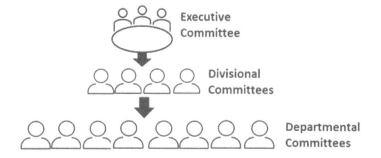

FIGURE 17.1 The safety committee system.

- Minutes of previous meeting.
- Outstanding matters.
- Reports:
  - Inspection feedback.
  - Accident investigation closeouts.
  - Accidents experienced.
  - Near miss incidents reported and action taken.
  - Health and safety representatives' reports.
  - Other.
- Items for discussion.
- Incident and accident recall.
- Action plan for the month.
- Close.

MEETING MINUTES

It is important to record the attendance at these meetings as well as the actions delegated to different members. These minutes serve as a record of the meeting and also a checklist for actions that need to be taken before the next meeting. Minutes from the departmental meeting would be circulated to the divisional committees, and the divisional committees' minutes would be shared with the executive committee. This immediately gives management a snapshot of what activities are occurring within health and safety throughout the organization.

A VITAL SMS COMPONENT

The safety committee system is a vital component of the SMS. A correctly structured safety committee system will ensure that there is joint participation and communication between management, the workforce, and all levels within the organization.

The divisional or departmental managers must chair the committees, as they are the level accountable for safety within those divisions or departments. The safety department can coordinate the meetings and be responsible for the administration of the committee, and for circulating the minutes of the meetings and preparing the necessary report for the meeting.

SUMMARY

Health and safety committees are formally constituted committees consisting of a cross-representation of management, employees, and worker representation. The process of establishing and maintaining committees can be measured on a monthly basis and be regarded as an important safety KPI.

To be effective, the committee meeting must be chaired by the manager of the department or division where the committee is constituted. Often the mistake is made of having the safety department head up the committee meeting which detracts from the purpose and effectiveness of the committee.

A correctly structured safety committee system will ensure that there is joint participation and communication between management, the workforce, and all levels within the organization.

18 Health and Safety Representatives Appointed

In his book, *Industrial Accident Prevention* published in 1932, H.W. Heinrich recommended that management appoints certain employees as *safety observers*. Their prime function was to assist management in maintaining a safe workplace by reporting hazards in the form of high-risk workplace conditions and actions to management for rectification.

This was one of the first innovations to get employees to participate and be involved in the health and safety movement.

HEALTH AND SAFETY REPRESENTATIVES

Health and safety representatives are employees who are either appointed, nominated or who volunteer to assist in the promotion of safety to ensure a safer work area for all. The representatives contribute to health and safety at the workplace and assist in the identification and elimination of hazards.

SAFETY MANAGEMENT PRINCIPLES

The appointment of health and safety representatives is based on the following safety management principles.

Principle of Safety Communication

The more employees are informed about the safety requirements and achievements, the more they are motivated to participate and accomplish safety results.

Principle of Safety Participation

Safety motivation increases in proportion to the amount of participation of the employees in the safety process. Safety is a common factor and safety activities should involve all employees. They should be informed of safety issues at all times and should be asked to give input and suggestions on aspects of safety, which directly or indirectly concern them.

Principle of Safety Recognition

Safety motivation increases as individuals are given recognition for their contribution to the safety effort. Commending and encouraging employees for safe actions goes far in ensuring that those safe actions are repeated. Good safety should be praised, and this praise should be made in public where possible.

WHO SHOULD BE APPOINTED?

Any employee can be selected as a health and safety representative, but in most cases, representatives are selected from and represent a specific work area. They are the employees most familiar with the machinery and the process of their workplace and are therefore ideal candidates to identify hazards within that area.

Safety representatives have often been referred to as the eyes and ears of management, as they are concerned with safety at the point of action. Although both management and employees are proportionately responsible for safety within an organization, management cannot always be present at the workplaces and therefore needs help. Safety representatives are the ideal employees to assist management in identifying and reporting both high-risk physical work conditions and high-risk behaviors. Other situations, which may lead to losses, can also be identified and reported by safety representatives.

TRAINING

The selected health and safety representatives should receive suitable instruction and training in the identification of hazards and how to complete the monthly inspection checklist. This training should be repeated annually.

DUTIES OF HEALTH AND SAFETY REPRESENTATIVES

The following functions and duties could be carried out by health and safety representatives:

- Inspection of the workplace, including any article, substance, plant, machinery, or health and safety equipment at that workplace monthly with a view to the health and safety of the employees.
- To present a written inspection report to management after each inspection.
- To attend meetings of the health and safety committee.
- To identify hazards which have potential to cause injuries.
- To constantly review the effectiveness of the health and safety measures in place within their area.
- To assist management to investigate and examine the causes of accidents.
- To investigate any complaints by employees concerning that person's health or safety.
- To make representations to management or the health and safety committee concerning hazards and threats to people's safety and health.
- To participate in inspections with local inspectors and to accompany inspectors and consultants doing inspections of the workplace.
- To attend formal accident inquiries.
- To present safety talks to employees.
- To assist in motivating employees to work safer.

FIGURE 18.1 The health and safety representative inspection system.

THE HEALTH AND SAFETY REPRESENTATIVE SYSTEM

On a regular basis, preferably monthly, appointed and trained health and safety representatives conduct an inspection of their work area using a checklist. They note hazards and other deviations from accepted safe practice on the checklist and once completed, they discuss the findings with their area manager or supervisor. The area manager or supervisor acknowledges the inspection report and takes action to rectify the hazards and deviations (Figure 18.1).

POSITIVE PERFORMANCE INDICATOR

The selection, training, and interaction of health and safety representatives is an ongoing, positive measurable management performance indicator. This indicator can be measured on a monthly basis by using the following criteria:

- Have adequate health and safety representatives been appointed?
- Has adequate training been provided?
- Have monthly inspections been carried out?
- Have the inspection reports been reviewed by the management of the area?
- Has action been taken to rectify the hazards reported?

SUMMARY

The appointment of health and safety representatives is an important program within the health and SMS. It is an ongoing process which should result in all areas within the organization being inspected each month by health and safety representatives. Health and safety representatives are the employees most familiar with the machinery and the process of their workplace and are therefore ideal candidates to identify hazards within that workplace. Their main duty is to inspect the workplace, including any article, substance, plant, machinery, or health and safety equipment at that workplace with a view to the health and safety of the employees. After the monthly inspection, they submit their inspection checklist to their area manager and discuss the finding. Management then takes action to rectify the hazards noted in the report. This is an ongoing process.

19 Safety Perceptions Surveys

A good measurement of safety management performance is the information derived concerning the health and safety culture (climate, attitude, and opinion) of the organization obtained by a formal safety perception survey. Sometimes referred to as a safety culture survey. A perception survey is a tool intended to drive continuous improvement and should not be considered a silver bullet.

SAFETY PERCEPTION SURVEYS

A safety perception survey provides a quantitative measure of how employees feel about current safety policies and procedures and provides them an opportunity to share their recommendations for improvement. It is facilitated by asking all employees to answer questions on a questionnaire openly and honestly. The questionnaire is anonymous, and no employee names are required. The results are then analyzed, and actions are taken to address the shortcomings in the health and safety systems.

HAZARD IDENTIFICATION

By using an employee safety perception survey, the organization is tapping its best resource for hazard identification. In addition, it has been suggested that the more employees are meaningfully engaged in the health and safety process, the more successful that process is.

IMPROVE HEALTH AND SAFETY

A safety perception survey allows an organization to better understand how its employees perceive health and safety and the organization's approach to health and safety management. It allows the organization to identify both strengths and weaknesses in the safety management processes, which enables it to continuously monitor and improve its approach to health and safety.

WHAT IS THE PURPOSE OF A SAFETY PERCEPTION SURVEY?

The safety perception survey is a workforce engagement tool which provides managers and supervisors with information about the overall safety climate, attitude or opinion of a team, department, and the organization as a whole. Such a survey is a tool to help an organization continuously improve health and safety efforts.

The survey can also be used to provide real-time feedback to the workforce and obtain their insights and suggestions on how to improve health and safety.

DOI: 10.1201/9781003357513-24

The survey provides focus areas for managers and supervisors to target and improve on. It offers a consistent assessment and specific insight at the work team level, assisting the organization to identify improvement opportunities, and develop action plans for these improvements.

Safety perception surveys are used in assessing the safety climate in an organization. The safety climate is influenced by both behavior and attitude. Safety perception surveys focus on attitudes and beliefs held by workers, supervisors, and management.

LEADING INDICATOR

A safety perception survey can act as a leading indicator to predict future accidents and downgrading events. A confidential survey allows employees to speak freely about what makes them feel unsafe at work, allowing for systems and controls to be adjusted, and actions to be assigned. This also provides insight into how safety communication has been received by different levels within the organization. Determining if the employees perceive that the safety system is flawed before a loss occurs, is proactive safety management. Research in this area suggests that when measured, perceptions can predict the likelihood of certain behaviors.

SAFETY PERCEPTION SURVEY OBJECTIVES

A good perception survey should:

- Evaluate the organization's perception of the safety management climate.
- Ask the same questions of managers and employees at different levels within the organization.
- Be easily and economically administered, analyzed, and evaluated.
- Provide managers with data in a format that allows for definitive conclusions to be made for corrective action decision-making.

AREAS TO COVER IN THE SURVEY

The following are the main areas that should be covered by the survey:

- Management commitment and involvement: This establishes that health and safety procedures are in place and that management is involved, committed, and supportive of the safety efforts.
- Involvement of employees: This determines if employees are involved at all levels and have access to all the information and resources needed.
- Hazard identification and control: This evaluates the system of hazard reporting and rectification and also employees' concerns concerning other hazards.
- Training: This evaluates the health and safety training that takes place and its effectiveness.
- Health and safety system effective analysis: This gives an overview of the implementation of health and safety processes and policies at ground level.

Safety Perceptions Surveys

Although the survey is intended to measure employee perceptions of health and safety, this broad base is better measured by identifying several more specific areas, such as:

- Employee perception of management's commitment to the safety system.
- Employee perceptions of their co-workers' commitment to the safety process.
- Employee perceptions of the effectiveness of safety training.
- Employee perceptions of their involvement in the organization's safety processes.

Assurances of confidentiality of all responses will go a long way to ensure honesty of the survey process.

SURVEY INSTRUMENT AND QUESTIONS

To ensure that the survey questions are valid, the organization must determine that they accurately measure what it is intended to measure. The questions must be posed in such a way as to obtain accurate and honest responses.

If the survey is not validated, then the results may not be useful and likely will result in resources being wasted on gaps or shortcomings that may or may not be reflective of the actual state of the organization's safety processes.

THE LIKERT SCALE

The Likert scale is a five-point scale that is used to allow the individual to express how much they agree or disagree with a particular statement. The Likert scale provides five possible answers to a statement or question that allows respondents to indicate their positive or negative strength of agreement or strength of feeling regarding the question or statement.

Common responses to questions used on the Likert scale type questionnaire are as follows:

1 = Strongly agree.
2 = Agree.
3 = Neutral.
4 = Disagree.
5 = Strongly disagree.

BENEFITS OF USING LIKERT SCALE QUESTIONNAIRES

The use of Likert scale type responses has certain advantages such as:

- User-friendly: Likert scales are closed ended and don't ask respondents to generate ideas or justify their opinions. This makes them quick to fill out and ensures they can easily yield data from large samples.

- Quantitative: Likert scales easily classify complex topics by breaking down abstract phenomena into recordable observations.
- Qualitative: Because Likert-type questions are not binary (yes/no, true/false, etc.), the company can get detailed insights into perceptions, opinions, and behaviors.

FEEDBACK

Good feedback from the survey provides employees with a general, nontechnical explanation of the findings, which include a summary of any changes to be implemented as a result of the findings, and a chance to express any concerns or questions about the survey process or the results. Giving employees a chance to have their views heard (and reflected back to them) helps improve employee buy-in to the safety perception survey process. The perception from the survey may be accurate even if it is difficult for management to hear.

REEVALUATION

The final element of the safety perception survey cycle is reevaluating the entire process once complete. What was learned and what remains unknown? What effects did the changes that were made (based on information gained from the initial survey) have on measurable outcomes? What improvements can be made to the health and safety system to make it more effective or efficient?

SUMMARY

A good measurement of safety management performance is the information derived concerning the health and safety culture (climate, attitude, and opinion) of the organization obtained by a formal safety perception survey. Sometimes referred to as a safety culture survey. A perception survey is a tool intended to drive continuous improvement and should not be considered a silver bullet.

A safety perception survey provides a quantitative measure of how employees feel about current safety policies and procedures and provides them an opportunity to share their recommendations for improvement. It is facilitated by asking all employees to answer questions on a questionnaire openly and honestly. A safety perception survey allows an organization to better understand how its employees perceive health and safety, and the organization's approach to health and safety management.

Good feedback from the survey provides employees with a general, nontechnical explanation of the findings, which include a summary of any changes to be implemented as a result of the findings, and a chance to express any concerns or questions about the survey process or the results.

20 Employees Trained in Health and Safety

A positive safety performance indicator is the number of employees who have received formal health and safety training during a specified period. Measurement periods could be monthly, six-monthly, or annually.

SAFETY MANAGEMENT PRINCIPLES

The safety management *principle of participation* states that motivation to accomplish results tends to increase as people are given opportunity to participate in matters affecting those results. Regular health and safety training gives employees an opportunity to participate in the safety programs and processes and motivates them to be a part of the process.

The safety management *principle of communication* states that motivation to accomplish results tends to increase as people are informed about matters affecting those results. The more employees are informed about health and safety issues, the more their willingness to become involved in the safety process.

Ongoing health and safety training not only informs and enlightens employees about hazards and how to work safely but also makes them willing participants in the health and safety management system (SMS).

ACCIDENT ROOT CAUSES

A lack of, or inadequate, training creates a lack of knowledge which is one of the root causes of accidents. To address this, a formal and structured program of health and safety training should form part of the SMS. This training should be more than the basic safety training required by local health and safety legislation.

INDUCTION TRAINING

Before an employee or contractor is allowed onto the work site, they should attend a compulsory safety induction training session. Induction training is given to new employees and contractors to ensure that they are familiar with the work environment, the risks within the workplace, and the specific work they will be carrying out.

The duration of this training varies from workplace to workplace dependent on the nature and hazard of the industry. Ideally, an eight-hour safety induction program will give the opportunity to thoroughly cover all the aspects of health and safety at that workplace. This is a requirement in high-risk industries and mines.

ANNUAL REFRESHER TRAINING

An annual refresher training period is a requirement of safety legislation in many countries. This means all employees must attend an annual full-day refresher training period where health and safety rules of the workplace are covered. Past accidents can be discussed and other important aspects of health and safety are reinforced.

FIRST AID TRAINING

Formal first aid training is important to ensure sufficient trained first aiders are available during all work shifts. Both cardiopulmonary resuscitation (CPR) courses and first aid training give employees the tools to respond to medical emergencies. However, these two types of training diverge where CPR courses focus on responding to cardiac arrest episodes via CPR, while first aid training incorporates a broader range of medical skills.

First Responder Training

Besides first aid and CPR training, first responders are taught to assess a medical emergency, make sure that the injured party has sufficient airway and ventilation in order to breathe, monitor vital signs, prepare, and move an injured party from an unsafe environment. They are also trained to apply a splint, control bleeding, and evaluate medical emergencies and other situations that are unsafe.

HEALTH AND SAFETY COORDINATORS

Ongoing health and safety training and professional certification for health and safety coordinators are important and would keep them updated concerning modern safety techniques such as:

- Risk management.
- Risk evaluation.
- Risk financing.
- Ergonomics.
- Sick building syndrome.
- Carpal tunnel syndrome, etc.

HAZARD COMMUNICATION TRAINING

Hazard communication, also known as HazCom, is a set of processes and procedures that employers and importers must implement in the workplace to effectively communicate hazards associated with chemicals and hazardous materials during handling, shipping, and any form of exposure.

CRITICAL TASK TRAINING

An important aspect in carrying out critical, or hazardous tasks, is the training that is given to enable the operator to understand the critical task procedure. Retraining after a job observation is the best method to ensure that operators follow the critical task procedures.

HEALTH AND SAFETY REPRESENTATIVE TRAINING

While initial and ongoing selection and training of health and safety representatives fall under the heading of health and safety training, it is usually measured as a separate key performance area.

COMPETENCY-BASED TRAINING

Competency- or outcome-based training refers to a learning model where students must demonstrate the required level of knowledge and skill (competency) on a task prior to advancing to the next task. It is a method of training that is focused on specific competencies or skills. Unlike other training methods, competency-based training is broken down into smaller units that are focused on one single key skill. The learner must demonstrate his or her mastery of that single skill or competency before continuing to the next segment of training. The skills are put together into modules and usually at the end, the learner receives some form of qualification or certification. This form of training is ideal for safety training that requires certain skills and competencies before the employee can be licensed or certified.

TECHNICAL HEALTH AND SAFETY TRAINING

Many aspects of health and safety training are technically orientated and competency based, and this could include the following:

- Vehicle safety.
- Occupational hygiene.
- Hazard identification.
- Near miss incident recognition and reporting.
- Accident investigation.
- Pressure vessel safety.
- Ladder safety.
- Permit issuance and receiving.
- Confined space team member.
- Lifting gear safety.
- Energy control (lockout).
- Trenching and shoring.
- Blood-borne pathogens.

- Understanding material safety data sheets (MSDS).
- Confined space entry.
- Asbestos awareness, etc.

FORMALIZED

The training provided should be guided by a syllabus and employees should sign an attendance sheet indicating that they have attended the training. Where appropriate, the necessary licenses or permits should be issued after the training so that employees can prove they are certified in certain aspects. Where needed, annual refresher training should form part of this training process.

ACCIDENT AND INCIDENT RECALL

Incident recall should form a part of all training programs where attendees can recall near miss incidents and accidents or other events from their past experience and share these with others attending the training.

KEY PERFORMANCE INDICATOR

Formal health and safety training should form part of the critical performance indicators allocated to managers at different levels. These indicators could include the number of employees trained in certain topics and progress could be monitored on a monthly or annual basis. An example is given in Figure 20.1.

SUMMARY

A positive safety performance indicator is the number of employees who have received formal health and safety training during a specified period. Measurement periods could be monthly, six-monthly, or annually.

Health and safety training should not be limited to legally prescribed training as per local health and safety regulations, but should be extended to ensure employees are fully informed about all aspects of health and safety and especially the critical processes that they will be involved in. The more employees are skilled in health and safety aspects, the more they will be able to partake in the safety processes and programs.

TYPE OF TRAINING	ANNUAL TARGET	ACTUAL	DEVIATION
First Aid Training	5% of the workforce annually		
Safety Induction	100% of new employees and contractors		
General Safety Training	50% of workforce annually		
Hazard Communication Training (HazCom)	100% of affected employees per year		
Critical Task Training	100% of affected employees per year		

FIGURE 20.1 An example of health and safety training key performance indicators.

The training provided should be guided by a syllabus and employees should sign an attendance sheet indicating that they have attended the training. Where appropriate, the necessary licenses or permits should be issued after the training so that employees can prove they are certified in certain aspects. Where needed, annual refresher training should form part of this training process.

21 Safety Toolbox Talks and Task Risk Assessments

Safety toolbox talks, also called tailgate meetings, a hard hat chat, or a health and safety brief, are gatherings of employees where a short message on health and safety is given by the supervisor or team leader. These talks are most commonly held on the job site, right before the start of the work shift. The talk acts as a reminder of the safety aspects of the job at hand and helps create awareness before or during the work task.

Task risk assessments are documented risk assessments of a task or work which is about to be undertaken. They are an assessment of what could go wrong during the process and what the possible consequence may be. Based on this information, identified hazards are rectified before the task commences and other suitable safety controls are implemented, as necessary.

PURPOSE OF TOOLBOX TALKS

The purpose of a toolbox talk is to convey a safety message to the work team and to create an awareness as to the hazards that could be encountered during their work. Specific instructions concerning the work at hand can be discussed and safety control measures can be emphasized during the talk.

As an example, toolbox safety talks of at least ten minutes in duration should be held in workplaces weekly, as a minimum. All employees must be exposed to a talk at least once per week. Office talks can be held on a monthly basis for office staff, exposing all office staff to at least one talk per month.

TOPICS

The topics and issues discussed at these toolbox talks should preferably be applicable and pertinent to the work being done and the workplace environment. Discussing general health and safety topics rather than focusing on the imminent hazards would be a waste of an opportunity to address site and job-specific risks.

OCCUPATIONAL SAFETY AND HEALTH ADMINISTRATION

The Occupational Safety and Health Administration (OSHA) has a series of Quick Cards available on their website which are one-page toolbox talk topics covering a wide range of health and safety topics. These are well-organized topics which are based on years of safety experience and give the dos and don'ts and other information on a wide variety of safety aspects.

Structured

The toolbox talk must be structured and follow a sequence and be guided by a written talk topic card. This sequence could begin with a welcome and introduction of the topic. The topic is then read with emphasis being placed on pertinent points. A conclusion should end the talk, after which attendees can ask questions and open discussions on the topic can take place.

Meaningful

The toolbox talk must be meaningful and cover a topic pertinent to the work at hand or the workplace environment. The time that is being used for the gathering is valuable production time and must be used to its best benefit and should not just be a "going through the motions" exercise.

Attendance Sheet

Employees attending the toolbox talk should sign the attendance register which could be printed on the back of the topic sheet. Attendance at these discussions is a proactive safety activity and can be included in the safety key performance indicators. Attendees should be encouraged to ask questions during the talk and incident recall should be encouraged.

TASK RISK ASSESSMENTS

An advancement on the toolbox talk is a task risk assessment (Figure 21.1). This is where the supervisor gathers the work team at the job site and asks the question, "What can go wrong during this task?" This question is discussed amongst the team to determine the probability of an accident occurring and then the question, "If it happens, how bad could it be?" is asked to determine the possible outcomes of the event if it occurs. The responses are plotted on the risk matrix on the task risk assessment card and the level of risk determined. Depending on this, the hazards are rectified, or the task does not continue until necessary safety measures have been taken.

Key Performance Indicator

The number of toolbox talks held, and the number of task risk assessments done should form part of a supervisor's key performance indicators. Targets for these proactive safety processes should be set based on past experience. Achievement should be monitored on a monthly basis. These activities are good positive leading indicators of safety effort.

As with all positive safety performance indicators, targets and objectives for these indicators should be SMART, meaning that they should be:

- Specific – the indicator must specify exactly what must be done in detail. It should not be vague or generalize.
- Measurable and manageable – the indicator must be measurable and manageable.

Daily Task Risk Assessment

Before each work task ask the following questions:
PROBABILITY= *What can happen here?* (Injury, damage, fire, etc.)
SEVERITY= *If it happens, how bad will it be?* (Death, injury, damage)
TASK..
Rank each question as Low (1) Medium (2) Medium-High (3) High (4)

Probability of Accident

	Low	Medium	Medium-high	High
High		8	12	16
Medium-high		6	9	12
Medium		4	6	8
Low				

How Bad Could it Be?

■ 12-16 Stop work immediately and fix the unsafe situation
4-9 Fix the unsafe situation and continue work
If the work is accepted as safe - proceed

Action taken: ....
....
Safe to work after action   YES [ ]   NO [ ]   (Stop Job)

Name........................Signature....................Date..........

FIGURE 21.1 An example of a daily or task risk assessment. (From McKinnon, R. C. 2017. *Risk-Based, Management-Led, Audit-Driven Safety Management Systems.* Boca Raton: Taylor and Francis. With permission.)

- Achievable and advantageous – the indicator must be achievable considering costs and resources. It should be aligned with the organization's objectives and be advantageous to the organization.
- Realistic and result oriented – performance indicators must be realistic, and results orientated.
- Time bound – the indicator must specify when, and how often the actions prescribed must be carried out.

SUMMARY

Safety toolbox talks, also called tailgate meetings, a hard hat chat, or a health and safety brief, are gatherings of employees where a short message on health and safety is given by the supervisor or team leader. Employees attending the toolbox talk should sign the attendance register which could be printed on the back of the topic sheet.

Task risk assessments are documented risk assessments of a task or work which is about to be undertaken. They are an assessment of what could go wrong during the process and what the possible consequence may be. Based on this information, identified hazards are rectified before the task commences, and other suitable safety controls are implemented, as necessary.

Attendance at toolbox talks, and the completion of task risk assessments, is a proactive safety activity and can be included in the safety key performance indicators.

22 Quality of Accident Investigation Reports

An accident is the result of failure in the management system. It could result in injury to employees, damage to property, or harm to the environment or a combination of consequences. While recording the losses after the event is a reactive action, lagging indicators do inform the organization about the types of losses being incurred, the frequency, and the cost thereof.

ACCIDENT INVESTIGATION

An accident offers management the opportunity to examine what went wrong, where there was a failure in the system, and what can be done to prevent a similar event or recurrence of the same event in the future.

An accident investigation offers management an opportunity to fix the problems that caused the accident. The problems can only be rectified if the investigation is done correctly and thoroughly, and the root causes of the event are identified. It is the deep-seated accident root causes that need to be rectified and not the obvious causes which would lead to treating the symptom and not the cause.

QUALITY OF INVESTIGATIONS

Since this is an opportunity to prevent future accidents, the importance of thorough accident investigations cannot be overemphasized. Although a reactive, after the event exercise, the accident investigation should produce action plans to implement improvements to prevent future undesired events. The quality of accident investigations is a leading indicator of safety performance. Each step of the investigation process can be evaluated, scored, and ranked.

The objective, therefore, would be to score 100% on each investigation report. This can also be built in as a key safety performance indicator and be measured on a monthly basis.

SCORING AN ACCIDENT INVESTIGATION REPORT

Once an accident investigation is completed and signed off, it should be submitted to the departmental manager for evaluation. Using a checklist, the manager allocates a score to each portion of the accident investigation form. Only by doing this quality check will management know how thorough and effective the accident investigation system is.

Quality of Accident Investigation Reports

EVALUATION CHECKLIST

A score sheet should be drawn up to correspond with the investigation form, and points be allocated to each segment, totaling 100 points.

GENERAL INFORMATION (10 POINTS)

The general information given on the form should be correct and accurate. This would include the correct names of the employees involved, the date and place of the event, the type of event, and a list of witnesses.

RISK ASSESSMENT (5 POINTS)

The risk assessment portion of the form should be correctly filled in, and should rank what *could* have happened, rather than what actually happened. Probability and severity are to be indicated.

DESCRIPTION OF THE EVENT (15 POINTS)

A brief, accurate, and factual description of the accident should be given. Although it may be difficult for the investigator to accurately determine the costs of the losses, some indication should be given as to the amount, even if it is an estimate. The investigator should indicate if the event was an injury, damage accident, a disruption causing event, or if it was something else such as a fire or accidental exposure. If a combination, this should be indicated.

To evaluate the written description of the event, the description and scoring could be:

- Comprehensive – 3 points.
- Clear – 3 points.
- Factual – 3 points.
- Accurate – 2 points.
- Detailed – 2 points.
- Event type specified – 2 points.

The scoring of this portion of the form should allocate points to each of the above criteria as indicated, using the score table (Figure 22.1).

EVENT DESCRIPTION	MAXIMUM SCORE	ACTUAL SCORE
Comprehensive	3	
Clear	3	
Factual	3	
Full description	2	
Detailed	2	
Event type	2	
TOTAL	15	

FIGURE 22.1 A scoring table for the description portion of the investigation form.

IMMEDIATE (CAUSE OF ENERGY EXCHANGE) CAUSE ANALYSIS (7 POINTS)

High-risk behaviors – all the high-risk behaviors involved in the event should be listed.

High-risk workplace conditions – all the high-risk workplace conditions noted at the accident scene should be listed.

ROOT CAUSE ANALYSIS (8 POINTS)

All the personal (human) factors relating to the immediate causes should be listed, as well as the job (workplace) factors. They must be derived from the immediate causes, and the investigator must be able to justify them. Simply writing down root causes because they seem applicable, is not acceptable. They must have been derived by a structured process and must relate to the immediate causes.

SKETCHES/PICTURES/DIAGRAMS

These should be clear and give a good idea of the accident site and losses as a result of the accident. They should also be accurate, factual, and relate directly to the event.

RISK CONTROL MEASURES (REMEDIES) (30 POINTS)

This is deemed to be the most important part of the investigation process, and management scrutinizing the accident investigation form will want to know if suitable, effective, and remedial measures to reduce the risks have been taken. This portion scores more than any other portion of the accident form. Are these recommended risk control measures feasible, cost-effective, and will they treat the root causes? Do they comprise an action plan that delegates tasks in line with the recommendations? Have these tasks been given a commencement and a completion date? Is follow-up action documented?

INVESTIGATION COMMENCEMENT (20 POINTS)

It is imperative to commence the accident investigation as soon as is practical after the event. Any delays could result in vital evidence being lost, being altered, or missed. Witnesses' recollection of the event also fades with time, so it is important to start the process timeously. Because of the importance of when the investigation was started, 20 points are allocated if the investigation started on the same day. Points are reduced for each day an investigation started after the event.

SIGNATURES (5 POINTS)

Correct signatures on the form are important to indicate the accident has been reviewed and the signatories agree with the findings. Final signatures are only allocated once

Quality of Accident Investigation Reports

all the recommendations for risk control have been implemented, and a follow-up has confirmed this. The form is then submitted to the next one up manager until it reaches the highest level of management within the organization (depending on the severity of the event). The sequence of signatures is as follows:

- Investigator.
- Supervisor.
- Manager.
- One-up manager.
- Executive.
- Safety department.

The safety department are the final signatories on the form. Their signature signifies that the details of the accident are accurate, that the root causes have been established, and that the risk control measures are appropriate and have been completely implemented. They also verify that all have signed off on the report form.

HAS THE ACCIDENT INVESTIGATION BEEN EFFECTIVE?

Once a final score has been allocated to the investigation report (Figure 22.2), management will have a good idea as to how thorough and effective the accident investigation has been. Low scores may indicate a need for retraining in accident investigation, or that the investigator did not take the task seriously enough. This measurement will indicate to management what action needs to be taken to improve the quality of accident investigations. It will also indicate which investigators do a thorough and meaningful investigation.

Item	Maximum Points	Actual Points
General information	10	
Risk assessment	5	
Description of the event	15	
Immediate cause analysis	7	
Root cause analysis	8	
Risk control measures	30	
Date of investigation	20	
Signatures	5	
TOTAL	100	

FIGURE 22.2 A scoring table to rank the quality of accident investigation reports.

SUMMARY

An accident is the result of failure in the management system. It could result in injury to employees, damage to property, or harm to the environment or a combination of consequences. An investigation after an accident offers management an opportunity to fix the problems that caused the accident.

Since this is an opportunity to prevent accidents, the importance of thorough accident investigations cannot be overemphasized. Although a reactive, after the event exercise, the accident investigation should produce action plans to implement improvements to prevent future undesired events.

The quality of accident investigation reports is a leading indicator of safety performance since each step of the investigation process can be evaluated, scored, and ranked. The objective, therefore, would be to score 100% on each investigation report. Once an accident investigation is completed and signed off, it should be submitted to the departmental manager for evaluation. Using a checklist, the manager allocates a score according to a scorecard to each portion of the accident investigation form. This is a key safety performance indicator and can be measured on a monthly basis.

Part VI

The Safety Management System (SMS) Audit as a Safety Management Performance Measurement Tool

23 The SMS Audit as a Safety Management Performance Measurement Tool

A health and safety management system (SMS) audit is carried out by an internal or external team and examines and quantifies all health and safety programs, processes, procedures, and risk mitigation activities of an organization. The audit assesses the current level of risk reduction, and subsequently helps in the preparation of an action plan to upgrade, modify, and improve health and safety inputs.

SAFETY MANAGEMENT PERFORMANCE MEASUREMENT

A comprehensive SMS audit is perhaps the best safety management performance measurement tool. The audit measures all health and safety activities within the organization and evaluates leading, as well as lagging indicators. The audit score is a measurement of a combination of leading and lagging indicators. It is a comprehensive and thorough evaluation of the SMS, and results in an accurate quantification of all aspects of health and safety. By comparing audit results, the organization can monitor its safety performance and implement action plans to address weaknesses and build on strengths. The audit period may differ from industry to industry, but the periods between the audits should not exceed 12 months and at a minimum, should be conducted every six months.

WHAT IS AN SMS AUDIT?

An SMS audit is a critical examination of all, or part, of a total SMS and is a management tool that measures the overall operating effectiveness of the SMS. An audit provides the means for a systematic examination, and analysis of each element, process, procedure, and program of a system, to determine the extent and quality of the controls. It confirms that management controls are in place to reduce risk, and that these controls are working. An audit measures the SMS effectiveness and highlights its strengths and weaknesses.

Not a Safety Inspection

An SMS audit is not simply a safety inspection, but rather a complete process of reviewing the SMS standards and their application, and effectiveness in the workplace. The audit does include an inspection which is the physical verification of the application of the standards in the workplace, and culminates in a review of the standards, policies, and programs during a formal documentation review session.

According to the Royal Society for the Prevention of Accidents (RoSPA) (2022), *Health and Safety Audits E-Book:*

> Safety audits are an essential part of a successful business. Effective health and safety auditing not only provides the legal framework for compliance, but it also lays the foundations for continuous safety improvement to enhance competitive advantage. The main duty of any health and safety auditor is to look at your organization's safety management systems and assess them in line with the chosen criteria.
>
> p. 5

PROACTIVE APPROACH

The audit is a proactive approach to measure health and safety performance of the organization. It helps in developing criteria for further improvements in organizational strengths along with the identification and control of organizational weaknesses. It provides a platform for taking effective planning decisions.

REACTIVE VERSUS PROACTIVE MEASUREMENT

Information on injury experience is reactive and is not control. Injury rates are lagging indicators and only measure the consequences of a weak or nonexistent SMS. Since they are measurements of consequence, largely dependent on fortuity and integrity of reporting, they do not accurately reflect risk reduction efforts. Since the majority of accidental events do not result in injuries, using their data as a sole safety gauge does not show the complete picture. SMS audits measure both leading and lagging indicators. They measure the amount of control an organization has over its risks. More meaningful information is obtained from systematic inspection, auditing of physical safeguards, systems of work, rules and procedures, and training methods, than on data about injury experience alone. Audits measure safety *effort* and injury rates measure safety *failure*.

SUBJECTIVE VERSUS OBJECTIVE

Audits of the SMS must not be subjective, which refers to personal perspectives, feelings, or opinions entering the decision-making process. This may be as a result of inexperience or some personal agenda with the organization being audited.

Audits should rather be objective, which is the elimination of subjective perspectives and a process that is purely based on hard facts. As with accident investigations, there should be fact-finding exercises and *not* fault-finding exercises.

BENEFITS OF SMS AUDITS

The concept of self-regulation is virtually accepted with the advent of performance standards and a duty of care approach. It is this obligation to self-regulate that reinforces the importance of an effective audit process as an essential management tool.

The SMS Audit as a Safety Management Performance Measurement Tool

Well-structured and conscientiously conducted audits provide an objective view of the actual SMS status. They identify weakness, recognize success, evaluate compliance, determine adequacy of standards, policies and procedures against statutory requirements, and organizational and world-class standards.

Further benefits of SMS audits include the following:

- The strengths, weaknesses, opportunities, and threats of the SMS are highlighted.
- The degree of conformance to regulatory requirement is measured.
- Compares what is being done against what was intended to be done.
- Evaluates whether the organization is achieving the goals of its safety standards.
- Helps in the prioritization of corrective actions.
- Assists in the compilation of a detailed report on the SMS.
- Compares the organizations SMS with world's best practice.
- Facilitates a continuous improvement process.

According to the Royal Society for the Prevention of Accidents (RoSPA) (2022), *Health and Safety Audits E-Book:*

> Whilst an audit is used to assess health and safety management systems, it is important to view an audit as a positive – it is a chance to highlight company successes and an opportunity to praise staff for their excellent work.
>
> p. 5

A Learning Opportunity

SMS audits are a learning opportunity and should be viewed positively. Sometimes the word "audit" conjures up an image of a painful experience and this should not be allowed to happen. The knowledge and experience of the auditors will be shared with the organizations' staff, and they will make a positive contribution to the health and safety efforts. They also have the advantage of coming from the outside and may see deviations, problems, or defects that employees in the workplace have become familiar with and accepted as the norm. (Can't see the wood for the trees.)

TYPES OF SMS AUDITS

Baseline Audit

This is more than likely the first formal audit an organization will experience. The audit of the existing health and safety system will be against legal requirements and world's best practice in health and safety. Regulatory requirements will be considered the minimum standard for this audit, and world's best practice health and safety standards will be the benchmark standard. Guidelines such as the International Organization for Standardization (ISO) ISO 45001:2018, American

National Standards Institute (ANSI) ANSI Z-10, Occupational Safety and Health Administration (OSHA), Voluntary Protection Program (VPP), and others can also be used to reflect world-class standards. Audit protocols are essential and substantial work is required to compile audit protocols to measure and quantify compliance with these and the organization's standards.

Records and Verification

The baseline audit can only be accurate if all pertinent documentation and records appertaining to health and safety systems in operation at the time of audit are available for scrutiny after the audit inspection. Employees who manage safety related issues should also be available to answer the auditor's questions.

Who Should Conduct the Baseline Audit?

The baseline audit should be carried out by a suitably experienced SMS auditor who has been exposed to world's best practice health and safety systems, and who has experience in auditing them. The auditor should ideally be accompanied by an assistant auditor who also has extensive knowledge of world's best practices in health and safety, and the auditing thereof.

Internal SMS Audits

Internal audits should take place every six months and be conducted by internal employees who have been trained in health and safety and in SMS auditing techniques. The organization must establish impartiality and objectivity of the audit by ensuring that the audit role is separated from the employee's normal function. In some instances, health and safety staff from neighboring organizations, or sister companies are invited to conduct these audits.

Independent Audit (External Third Party)

Independent audits, or external third-party audits, should take place annually. They should be conducted by expert agencies comprising well-trained and experienced lead auditors, technical safety experts, SMS audit specialists, and others. Organizations must give them full freedom to judge neutrally and independently without exercising any influence from any quarter. Objectivity is vital, and their disclosures are to be taken as bare facts, based on which further lines of correction and improvements follow.

Compliance Audit

A compliance audit is an audit to measure compliance with local regulatory requirements. The main question to be asked during a compliance audit is, "Does the organization comply with the requirements of the legal health and safety regulations?"

CENTRALIZED AND DECENTRALIZED SMS COORDINATION

In an organization with different departments that are physically decentralized from the head office, those departments would manage their own SMS on a decentralized basis. Head office would maintain a centralized version with elements that cover the entire organization. Together, both would comprise the company SMS.

Centralized elements, for example, would include the following:

- Health and safety policy.
- Health and safety newsletters.
- Employee pre-employment medicals.
- Company safety goals.
- Executive health and safety committee, etc.

Decentralized elements, for example, would include the following:

- Housekeeping.
- Permit to work systems.
- Appointment of health and safety representatives.
- Safety inspections.
- Motorized transport safety, etc.

AUDITABLE ORGANIZATION

Before any SMS audit, it should be determined if the organization represents an auditable unit. To qualify as such, the organization to be audited should be geographically, organizationally, and operationally together. The organization should be able to sustain and maintain a comprehensive 70 or more element safety system. Outlying offices, workplaces, depots, and other worksites are included in the audit and form part of the organization for audit purposes.

A comprehensive audit of a warehouse (as an example) that maintains only a few elements of the SMS, as they are the only elements applicable and required due to the nature of the activities of the warehouse, is not feasible. The disadvantage of this is that if they are audited against fewer elements and the larger divisions are audited against the complete SMS, comparisons of scores are unrealistic and inaccurate. This unfair comparison could lead to upsets.

Solution

A possible solution is to have different levels of audit protocols. For example:

- A main 70+ element SMS audit protocol for the main organization.
- A 40-element audit protocol for smaller divisions that maintain a smaller scale SMS, such as an autonomous branch away from the main organization.
- An office audit protocol for distant offices that are not geographically close to the main organization. The protocol is designed for offices and contains the relevant applicable elements.

These three levels of audit protocol will enable each workplace to be audited against applicable standards. All workplaces should be audited using the elements that are applicable to them.

AUDIT REQUIREMENTS

The audit can only quantify the effort during the preceding 12 months before the audit. The audit cannot measure future improvements, innovations, or amendments. It can only measure what was done and achieved during the last year. No recognition or score can be given for future intentions.

Elements, processes, and programs that have not been in operation for at least six months cannot be considered for a 100% score. These processes must have been in place, and operating, for at least six months to be considered should they qualify for full marks.

REVIEW OF ELEMENT STANDARDS

A copy of each element standard should be available for the auditors to review. Element standards and their verification documents are normally kept in a file, or folder, which is opened for each element. This file could be in electronic form. What is important is that all information pertaining to the element is stored in the folder, or file, and updated as the SMS progresses.

DOCUMENT CONTROL SYSTEM

A document control system, which is an element of the SMS, must ensure that updated and revised documents replace the old, outdated documents as soon as possible, and all employees are informed of the update or change.

INCOMPLETE VERIFICATION

Verification documents that are not available or not signed where required will indicate a system breakdown. The rule of no paper, no points will apply. Incorrectly completed checklists or risk assessments should be identified as weaknesses in the system as well. The auditors must ensure that the verification matches what is happening in the workplace. If a system was followed during the inspection, the relevant documentation should be requested and checked. The main question auditors should ask is, "Is there a process in place, and is it working?"

The audit protocol will guide the auditor as to what questions to ask and what verification to call for to verify that the actions are taking place.

AUDIT FREQUENCY

Both the frequency and extent of the audit will be based on the complexity and level of maturity of the SMS. Changes in the workplace, legal requirements, or an increase in work-related injuries and illnesses may call for more frequent audits.

VERIFICATION

Audits establish the adequacy of the SMS and ascertain if it has been implemented appropriately, in relation to the risk and nature of the business. Audits determine whether or not the controls being applied are suitable, and how they fit the organization, its operations, and existing culture. Most importantly, the audits measure the effectiveness of the SMS and its contribution to the achievement of stated health and safety goals.

SMS AUDIT PROTOCOL

An audit protocol is a vital audit tool. It guides SMS auditors and helps them audit against a specific standard. It facilitates comparing the company against its own standards. The audit protocol should be compiled to cover the company SMS standards. This means it summarizes the requirements of the standards into measurable and auditable elements, minimum standards, and minimum standard details. It also establishes an element risk ranking dependent on the degree of risks that elements, processes, or programs pose, or their contribution to risk reduction.

THE AUDIT PROTOCOL

The protocol identifies:

- What must be looked for during the inspection?
- What questions to be asked during the system verification meeting?
- What documents, or verification, need to be reviewed? (Figure 23.1).

It indicates what elements must be followed through the system, and the processes that need to be examined. The protocol allocates a score for each element of the SMS and each minimum standard detail for scoring purposes. The audit protocol ensures that the auditors evaluate and measure each and every element of the SMS, and that each detail is scrutinized, and each requirement ranked or scored.

WHO SHOULD CONDUCT AUDITS?

Experienced and qualified auditors are essential if the audit process is to be successful. Auditing a workplace and exposing its controls for evaluation is not an easy thing to do and could lead to some supervisors being sensitive to criticism. The auditors must be professional at all times and ensure that the employees in the areas being audited see it as part of a learning experience. Compliments must be paid wherever possible. Leading questions should be asked, and the shop floor employee should be made to be a part of the audit, not a subject of the audit.

AUDITOR'S TRAINING

While it is difficult to prescribe exactly what training and experience an auditor should have, it is imperative that the person has a health and safety background and

ELEMENT / PROGRAM / PROCESS	POINTS	QUESTIONS THAT COULD BE ASKED	VERIFICATION	WHAT TO LOOK FOR
SAFETY INSPECTION SYSTEM				
Risk matrix used	5	Is a risk assessment method used to determine inspection frequency?	Copy of risk assessment	High-risk areas/processes
MONTHLY INSPECTIONS BY HEALTH AND SAFETY REPRESENTATIVES				
Health and safety representatives complete a monthly checklist based on the SMS elements	5	Do health and safe representatives or supervisors conduct regular, monthly safety inspections of their work areas? Are the checklists up to date and done correctly?	Copy of checklist	Monthly inspections by health and safety representatives/supervisors of their own areas and reports being submitted
Inspections to cover the entire area for which they are responsible	5	Are all areas covered? Is a checklist used?	Check at least 8 items from the SMS	Is it obvious that a health and safety representative or supervisor has inspected the area and taken action to rectify hazards?
High-risk conditions reported	5	Does the checklist include conditions?	Check some of the health and safety representative's checklists	Checklist based on SMS being used?
High-risk behaviors reported	5	Does the checklist include unsafe practices?	Are they on the form?	Review inspection checklist
Positive recommendations made	5	Are positive actions taken?	Example of actions taken	
Hazard training	5	Has the health and safety representatives been made aware of all the hazards in the area by the departmental supervisors?	Training/on-site training done	
REPORTS CONSIDERED AND ACTIONED BY MANAGEMENT				
All reports forwarded to management (assigned person or deputy?)	5	Does management review these checklists and authorize necessary action?	Check some reports and see where management has authorized action if necessary	Reports considered and actioned by management
Remedial action	5	Does management take action on the report findings and recommendations? How is this done?	Call for the checklist of that area and crosscheck	Note one area that has basic deviations and follow through the system
Feedback given to the health and safety representatives who submit the reports?	5	How and when is this done?	Refer to copies of report	
BI-ANNUAL INSPECTION BY SAFETY DEPARTMENT				
Health and safety department conduct bi-annual inspections of the entire worksite	5	Do the health and safety coordinators conduct inspections of the entire premises?	Look at checklist/inspection form used	
Other inspections	5	What other safety inspections take place?	Examples	
Findings of the inspection reported to management. Inspector follow-up on all inspection reports	5	Are the findings reported to management? Is there a follow-up?	Example of a follow-up after an inspection	
TOTAL	**65**			

FIGURE 23.1 An example of an audit protocol for the SMS element *safety inspection system* showing the minimum standard detail, the questions to be asked, the documents to be reviewed, and what to look for during the audit inspection. (From McKinnon, R. C. 2020. *The Design, Implementation and Audit of Safety Management Systems* (Figure 32.3). Boca Raton: Taylor and Francis. With permission.)

has been trained in auditing techniques. As a minimum, over and above the person's professional training, the following classes would be recommended:

- The basics of safety management – 8 hours.
- Practical risk management – 8 hours.
- Accident and near miss incident reporting and investigation – 8 hours.
- Critical task identification and procedure writing – 8 hours.
- Legal requirements – 6 hours.
- Internal accredited auditor's workshop – 40 hours.

Once these classes have been attended, the aspiring auditor should accompany a qualified auditor on at least three audits. Depending on how the aspirant auditor performs during these audits, he or she could then be accepted as an internal accredited auditor.

GUIDELINES FOR AUDITORS

The audit should be regarded as a learning experience. Criticism should be avoided, and a positive approach should be taken. The auditor must be factual and not find fault. Professionalism is necessary. The organization is exposing its workplaces and employees to scrutiny and the experience must result in a rewarding experience for them. The auditor must always pay compliments where possible and offer positive advice and recommendations.

AUDITOR'S EXPERIENCE

While there are formal training programs for both internal and external auditors, years of experience is vital before an auditor can be fully effective. The auditor should know their subject thoroughly and be able to impart his or her knowledge in a professional way.

The auditor should have experience in implementing, as well as auditing of an SMS. Knowledge of the workings of SMS elements is important, as is knowledge and understanding of management principles and practices. Being an auditor, one would expect the auditor to be a professional member of a health and safety body and hold acceptable certifications.

Internal audits can be very political and extra caution is required to select and train the correct people as auditors, and to ensure they remain unbiased at all times.

LEAD AUDITOR

The lead auditor is an auditor who leads the audit and is assisted by other auditors. The lead auditor should have extensive audit experience and must be able to conduct the audit in a professional manner.

QUESTIONING TECHNIQUES

Questioning techniques are important to derive as much information as possible in the short time allocated. Questions should be asked during the inspection and auditors should do most of the listening.

COMPLIMENT

To keep the audit on a positive note, auditors should complement adherence to standards where possible. Areas in the workplace that exhibit good housekeeping should receive praise.

SUMMARY

The best method to measure safety management performance is via a thorough and comprehensive SMS audit. This process will examine all aspects of the system and evaluate the status of each element, program, system, and process of the SMS against the standards set by local safety and health laws (as a minimum) and the organization's own health and safety standards.

An audit results in an accurate quantification of all aspects of health and safety. By comparing audit results, the organization can monitor its safety performance and implement action plans to address weaknesses and build on strengths. Auditing is a proactive approach to measure health and safety performance of the organization, and it provides a platform for taking effective planning decisions.

Audits should rather be objective, which is the elimination of subjective perspectives and a process that is purely based on hard facts. As with accident investigations, there should be fact-finding exercises and not fault-finding exercises.

An audit protocol guides SMS auditors and helps them audit against a specific standard. It facilitates comparing the company against its own standards. The audit protocol should be compiled to cover the company SMS standards. This means it summarizes the requirements of the standards into measurable and auditable elements, minimum standards, and minimum standard details. It also establishes an element risk ranking dependent on the degree of risks that elements, processes, or programs pose, or their contribution to risk reduction.

24 Example Leading Safety Key Performance Indicators

The following is an example of a health and safety management system (SMS) standard for leading safety key performance indicators (KPIs).

OBJECTIVE

The objective of this standard is to ensure that upstream, proactive safety, and health activities are used to measure safety performance, and to move the focus from lagging, reactive indicators such as injury and fatality rates, to more meaningful measurable management performances.

RESPONSIBILITY AND ACCOUNTABILITY

- Each area manager, supervisor, and contractor (responsible person [RP]) will be accountable for the compliance of this standard in their work area.
- All departments throughout the company and its subsidiaries will apply this standard and the new safety performance indicators.

REQUIREMENTS

The table (Figure 24.1) is to be used as a ranking table to allocate scores for the various criteria mentioned later. The maximum score is 100 for the ten key performance areas. The minimum measurement period is one week or in some instances, one month.

SAFETY COMMITTEES

The number of safety committee meetings relates to formal health and safety committees established as per company standards and which meet once per month for an hour as a minimum. Managers are to lead the safety committee meetings as the chairperson of the committee. Committees can meet more frequent but once a month is the minimum standard.

ACTIVITY	VERIFICATION	MAXIMUM SCORE	ACTUAL SCORE
Number of safety committee meetings held	Copies of the minutes of meetings held	20	
Number of near miss incidents reported	Number of written near miss incidents reported during period (copies)	10	
Safety observations acted on and completed	Number of, and completed safety observations (copies)	10	
Plant inspections completed	Copies of completed inspection checklists	10	
Safety system audit score	Internal audit score sheet	5	
Fire or evacuation drills held	Number of documented fire or emergency drills held	5	
Safety toolbox talks held	Copies of signed attendance sheets	10	
Employees attending safety and health training	Copy of attendance sheets	5	
Number of health and safety representatives appointed and active	Copies of appointment letters and inspection reports	15	
Daily or task risk assessments carried out	Copies of completed risk assessments	10	
	SCORE	100	

FIGURE 24.1 A score sheet for ten key performance indicators (KPIs) showing the criteria, the verification, and maximum score.

Scoring Method (Maximum 20 Points)

- A formal safety committee is established in the Department/Division = 5 points.
- The committee meets monthly = 5 points.
- The committee is chaired by management = 5 points.
- Minutes of the meetings are kept and are available = 5 points.

Alternate Score

- The committee meets 6 times per year (Bi-monthly) = 2.5 points.
- The committee meets twice per year (Bi-annually) = 1 point.

NEAR MISS INCIDENT REPORTS

The number of near miss incident reports received is measured by the number of near miss incident reports received from employees and contractors and should be equal to 8% of employees reporting at least one near miss per month. This equates to approximately a minimum of one near miss incident report per employee per year. Reports must be in writing, must be recorded, and follow-up action must be documented.

Scoring Method (Maximum 10 Points)

- 8% of workforce submitted near miss incident reports per month = 5 points.
- 100% of near miss incident reports have been acted on per month = 5 points.

Alternate Score

Reports:

- 4% of workforce submitted near miss incident reports per month = 2.5 points.
- 1% of workforce submitted near miss incident reports per month = 1 point.

Action taken:

- 50% of near miss incident reports have been acted on per month = 2.5 points.
- 20% of near miss incident reports have been acted on per month = 1 point.

SAFETY OBSERVATIONS REPORTED

Safety observations include the following:

- Reports of high-risk acts.
- Reports of high-risk conditions.
- Safe behavior reports.
- Informal inspection reports.
- Safety representatives' casual observations.

Safety observations are employee observations and feedback on safety issues noted by them. Once reported, some form of action must follow to either rectify the situation, act, or acknowledge the occurrence.

The number of safety observations reported and acted on should be equal to 10% of employees reporting safety observations that are acted upon each month. If there are 100 employees in the department, the indicator should be ten safety observations reported and acted upon each month, or 120 observations per annum, as a minimum.

Scoring Method (Maximum 10 Points)

- 10% of workforce submitted observations per month = 5 points.
- 100% of observations reported have been acted on per month = 5 points.

Alternate Score

Submission:

- 5% of workforce submitted observations per month = 2.5 points.
- 2% of workforce submitted observations per month = 1 point.

Action taken:

- 50% of observations reported have been acted on per month = 2.5 points.
- 20% of observations reported have been acted on per month = 1 point.

PLANT (WORKPLACE) INSPECTIONS COMPLETED

These are inspections which exclude formal inspections by health and safety representatives and safety staff.

Workplace inspections should be conducted at least monthly and more frequently based on risk assessments. This indicator will measure the actual number of inspections in relation to the frequency dictated by the risk assessment.

If the assessment proposes monthly workplace inspections, then the criterion is one inspection per month or 12 per annum. If it proposes weekly inspections, then the criterion is one inspection per week or 52 per annum.

Scoring Method (Maximum 10 Points)

- Weekly safety inspections conducted (4 per month) = 10 points or, dependent on the risk assessment recommendations.
- Monthly safety inspections conducted (12 per annum) = 10 points.

Alternate Score

- Safety inspections conducted less frequent than weekly (<4 per week) = 5 points.
- Safety inspections conducted less frequently than monthly (<12 per annum) = 5 points.

SAFETY MANAGEMENT SYSTEM AUDIT RESULTS

The score given for this indicator will be directly linked to the score awarded to the division or department during its last internal SMS audit. If the department or division has been assessed at 90% or more, then it would score 5 points for this indicator, if it scored 80% or more, it would score 4 points, etc.

Example Leading Safety Key Performance Indicators   211

SCORING METHOD (MAXIMUM 5 POINTS)

- Audit score 90% plus = 5 points.
- Audit score 80% plus = 4 points.
- Audit score 60% plus = 3 points.
- Audit score 40% plus = 2 points.
- Audit score 20% plus = 1 point.

FIRE, EVACUATION, OR EMERGENCY DRILLS HELD

A fire drill (evacuation or emergency drill) should be held at least every year and more frequently dependent on the degree of risk in the particular work area. The indicator is at least one evacuation drill per year per department or division. If the risk of the area dictates more frequent evacuation exercises, that will be the measurable number.

SCORING METHOD (MAXIMUM 5 POINTS)

- One evacuation drill held per year = 5 points.
- No drills held = 0 points.

NUMBER OF SAFETY TOOLBOX TALKS HELD

Toolbox safety talks of at least ten minutes in duration should be held in workplaces weekly as a minimum. At least 10% of employees must be exposed to a talk once per week. Talks for office staff can be held on a monthly basis, exposing all office staff to at least one talk per month.

SCORING METHOD (MAXIMUM 10 POINTS)

- One toolbox talk per week or 4 per month = 5 points.
- 10% of staff attended formal weekly toolbox talks = 5 points.

ALTERNATE SCORE

Talks held:

- 25 Toolbox talks held per year (one per fortnight) = 2.5 points.
- 12 Toolbox talks held per year (one per month) = 1 point.

Attendance:

- 5% of staff attended formal weekly toolbox talks = 2.5 points.

EMPLOYEES ATTENDING HEALTH AND SAFETY TRAINING PROGRAMS OR WORKSHOPS

At least 10% of employees must attend formal safety and health training courses per year. These courses must be formal, internal, or external courses of a minimum duration of eight hours. This figure excludes training attended by safety representatives, and training required by safety legislation.

Scoring Method (Maximum 5 Points)

- 10% of staff attended formal safety training = 5 points.
- 5% of staff attended formal safety training = 2.5 points.
- 2% of staff attended formal safety training = 1 point.

NUMBER OF SAFETY REPRESENTATIVES APPOINTED AND ACTIVE

At least 5% of the total workforce of a department or division must be appointed as health and safety representatives. This appointment must be done in writing and the safety representative must have attended the safety representative training program.

This index will measure the percentage of safety representatives appointed as well as the monthly inspection reports submitted and acted on by management.

Scoring Method (Maximum 15 Points)

- 5% of workforce appointed per current year = 5 points.
- Correct number of monthly reports (5% × workforce) = 5 points.
- 100% of reports signed and follow-up action completed = 5 points.

Alternate Scores

Appointments:

- 2.5% of workforce appointed per current year = 2.5 points (2.5% × workforce) = 2.5 points.

Follow-up completed:

- 50% of reports signed and follow-up completed = 2.5 points.

TASK OR SITE RISK ASSESSMENTS CONDUCTED

Before tasks or assignments are carried out, a daily on-site risk assessment should be conducted using a suitable checklist which incorporates a risk matrix. These could be assessments of new tasks, unusual work, permit required, or other tasks.

Scoring Method (Maximum 10 Points)

- Daily onsite risk assessment conducted (5 per week, 20 per month) = 5 points.
- Checklist with risk matrix used and completed = 5 points.

SCORE SHEET

Figure 24.1 shows the score sheet for the ten KPIs. It lists the activity, the verification of the activity, and the maximum scores for each activity.

SUMMARY

Proactive leading safety KPIs can be introduced to monitor safety management and assess performance. These indicators are elements of the total health and SMS and can be measured on a monthly basis.

The number of near miss incident reports received is measured by the number of near miss incident reports received from employees and contractors and should be equal to 8% of employees reporting at least one near miss per month.

Safety observations are employee observations and feedback on safety issues noted by them. Once reported, some form of action must follow to either rectify the situation, act, or acknowledge the occurrence. Workplace inspections should be conducted at least monthly and more frequently based on risk assessments.

The standard is at least one evacuation drill per year per department or division. If the risk of the area dictates more frequent evacuation exercises, that will be the measurable number.

Toolbox safety talks of at least ten minutes in duration should be held in workplaces weekly as a minimum. At least 10% of employees must be exposed to a talk once per week.

At least 10% of employees must attend formal safety and health training courses per year and at least 5% of the total workforce of a department or division must be appointed as health and safety representatives.

Before tasks or assignments are carried out, a daily on-site risk assessment should be conducted using a suitable checklist which incorporates a risk matrix.

References

Austin American Statesman. 2022. *DuPont Must End Its Safety Charade* (statesman.com.).
Ball, Lestie. 1970. RLS Human Care. Energy Release Theory of Accident | RLS HUMAN CARE (rlsdhamal.com)
Bird, F. E. Jr. and Germain, G. L. 1992. *Practical Loss Control Leadership* (2nd ed.). Loganville, Georgia: International Loss Control Institute.
British Safety Council. 1974/1975. *Tye–Pearson Theory.* Tye-Pearson Accident Pyramid | PDF | Traffic Collision | Accident (General) (scribd.com)
DNV (www.dnv.com) 2023. *A Tribute to Frank E. Bird Jr. (1921-2007)* – DNV.
E&MJ Engineering and Mining Journal. 2023. *Sibanye-Stillwater's South Africa Gold Operations Achieve 10M Fatality-free Shifts* | E & MJ (e-mj.com).
International Risk Management Institute. 2023. Haddon's energy release theory. Energy release theory (irmi.com)
Government Accountability Organization (GAO) (US). 2021. *GAO-21-122, Workplace Safety and Health: Actions Needed to Improve Reporting of Summary Injury and Illness Data.*
Guinness World Records 2023. *Most Zero Lost Time in Man Hours.* | Guinness World Records (https://www.guinnessworldrecords.com/world-records/zero-lost-time-%E2%80%93-most-man-hours).
Haddon, W. 2023. *Haddon's Countermeasures – A Better Alternative to the Hierarchy of Hazard Control?* (safetyrisk.net).
Health and Safety Executive (HSE) UK. 2006. *An Investigation of Trends in Under-reporting of Major and Over-3-Day Injuries in the Manufacturing Sector: First Survey* (hse.gov.uk.). (Contains public sector information published by the Health and Safety Executive and licensed under the Open Government License v1.0.)
Health and Safety Executive (HSE) UK. 2020. *Key Figures for Great Britain (2020/21).* Health and Safety Statistics (hse.gov.uk.) (Contains public sector information published by the Health and Safety Executive and licensed under the Open Government License v1.0.) Health and safety statistics (hse.gov.uk).
Health and Safety Executive (HSE) UK. 2022. *A Guide to Measuring Health and Safety Performance.* Managing Health and Safety Performance (hse.gov.uk.) (Contains public sector information published by the Health and Safety Executive and licensed under the Open Government License v1.0.) Guide to measuring health and safety performance (polfed.org).
Heinrich, H. W. et al, 1959. *Industrial Accident Prevention* (4th ed.). New York: McGraw-Hill Book Company.
Hoyle, B. 2005. *Fixing the Workplace and Not the Worker. Fixing the Workplace, Not the Worker: A Worker's Guide to Accident Prevention.* Lakewood: CO. Oil, Chemical and Atomic Workers International Union.
International Labor Organization. (ILO). 2000. *ILC 90 – Report V(1) – Recording and Notification of Occupational Accidents and Diseases and ILO List of Occupational diseases.* (ILO website, Copyright © International Labor Organization 2000).
International Labor Organization. (ILO). 2023a. *World Statistics. World Statistic.* (ilo.org.) (ILO). website, Copyright © International Labor Organization 2023).
International Labor Organization (ILO). 2003b. *Safety in Numbers.* Acknowledgements (ilo.org) (ILO website, Copyright © International Labor Organization 2023).

McKinnon, R. C. 2000. *The Cause, Effect, and Control of Accidental Loss, With Accident Investigation Kit. (CECAL)*. 6000 Broken Sound Parkway NW, Suite 300, Boca Raton, FL: CRC Press, Taylor and Francis Group.

McKinnon, R. C. 2007. *Changing Safety's Paradigms*. Maryland: Bernan Press.

McKinnon, R. C. 2012. *Safety Management, Near Miss Identification, Recognition, and Investigation* (Model 2.2). 6000 Broken Sound Parkway NW, Suite 300, Boca Raton, FL: CRC Press, Taylor and Francis Group.

McKinnon, R. C. 2017. *Risk-based, Management-led, Audit-driven Safety Management Systems*. 6000 Broken Sound Parkway NW, Suite 300, Boca Raton, FL: CRC Press, Taylor and Francis Group.

Mckinnon, R. C. 2020. *The Design, Implementation and Audit of Safety Management Systems*. 6000 Broken Sound Parkway NW, Suite 300, Boca Raton, FL: CRC Press, Taylor and Francis Group.

Mine Safety and Health Administration (MSHA). 2010. *Final Report – Fatality #3 through #31 – April 5, 2010* | Mine Safety and Health Administration (MSHA).

National Safety Council (US). *Injury Facts* 2022. Permission to reprint/use granted by the National Safety Council © 2022. Website 2023. Work Safety Introduction – Injury Facts (nsc.org).

Occupational Safety and Health Administration (OSHA). 2015. *Near Miss Report Form* (osha.gov).

Occupational Safety and Health Administration (OSHA). 2023. Top 10 Most Frequently Cited Standards | Occupational Safety and Health Administration (osha.gov).

Petersen, D. 1978. *Techniques of Safety Management* (2nd ed.). New York: McGraw-Hill Book Company.

Petersen, D. 1998. *What Measurement Should We Use, and Why? Professional Safety*, October, p.37.

Rothstein, M. A. 1998. *Occupational Safety and Health Law* (4th ed.). St. Paul: West Group. Reproduced with permission.

Royal Society for the Prevention of Accidents (RoSPA) (UK). 2022. *Health and Safety Audits e-Book*. PowerPoint Presentation. (rospa.com.) (Contains public sector information licensed under the Open Government License v3.0.)

Royal Society for the Prevention of Accidents (RoSPA) (UK). 2023. *Measuring OS&H performance – RoSPA*. (Contains public sector information licensed under the Open Government License v3.0.)

Safeopedia.com. 2022. *What Is an Accident?* – Definition from Safeopedia.

Tarrants, W. E. 1980. *The Measurement of Safety Performance*. New York: Garland STPM Press.

The Guardian. 2022. *Revealed: 6,500 Migrant Workers Have Died in Qatar Since World Cup Awarded* | Workers' rights. | The Guardian.

US Bureau of Labor Statistics. 2023. *Injuries, Illnesses, and Fatalities (bls.gov.)*

US Government Publishing Office, House Hearing, 110 Congress. 2008. *Hidden Tragedy: Underreporting of Workplace Injuries and Illnesses* (govinfo.gov.).

Index

Note: Page numbers in *italics* refer to figures

A

Accident
    causation theories, 14
    clarity, 13
    definitions, 11, 12
    domino effect, 14
    injury, 13–14
Accidental loss causation
    cost of accidental loss, 31, *31*
    exposure, impact, or exchanges of energy, *30*, 30–31
    fortuity, 29–30
    high-risk (unsafe) behavior, 28, 29
    high-risk (unsafe) conditions, 28, 29
    HIRA sequence, 27, *27*
    loss causation sequence, 25
    losses, 26
    near miss incident, 26–27
    root causes of, 27–28, *28*
    sequence of events, 25
    slightly different circumstances, 30
    warnings, 30
    weak/non-existent SMS, 27, *28*
Accident and near miss incident, 52–53
Accident costs
    cost–benefit analysis, 117
    cost of non-compliance, 115–116
    cost of risks, 118
    cost reduction, 114–115
    experience modification rate (EMR), 111–112
    fines, 115
    hidden costs, 112–113
    iceberg effect, 113, *114*
    incidental costs, 112
    increased fines and penalties, 116
    increased premiums, 111
    life's value, 117
    major accidents, 116–117
    minimizing losses, 113
    profit-driven, 114
    reputation, 117
    severe repercussions, 117
    totally hidden costs, 113
    white paper, 118
Accident investigation reports
    evaluation checklist, 191–193
    general information, 191
    immediate cause analysis, 192
    investigation commencement, 192
    quality of investigations, 190
    risk assessment, 191
    risk control measures, 192
    root cause analysis, 192
    scoring table, 193, *193*
    signatures, 192–193
    sketches/pictures/diagrams, 192
Accident ratios
    amnesty, 100
    calculation, 99–102
    costing, 102
    criticism of, 97–98
    downgrading events, 99–100
    health and safety executive (HSE), *97*
    high-risk behaviors, 145
    high-risk workplace condition, 145
    injury-resulting accidents, 95, *96*
    investigation and recording, 98
    large numbers, 135
    near miss incident, 134–135
    organization's loss figures, *101*
    risk matrix, *98*, 98–99
    study, 96
    Tye–Pearson theory, 96–97, *97*
    vital statistics, 100
Agency
    agency part, 75, 108
    analysis, *108*
    general agencies, 74
    occupational hygiene, 74
    principal object, 107
    trend analysis, 75, 108
    types, 73–74, *74*
Aircraft accidents, 76
ALARP (As Low as Is Reasonably Practicable) zone, 39, 166
American National Standards Institute (ANSI), 59, 68–69, 200
Awards, lagging indicators, 89–93
    Mining Awards, 91
    National Occupational Safety Association (NOSA) Awards, 89–90
    National Safety Council (NSC) (US) Awards, 90–91
    NOSA 5-Star Safety and Health Management System, 91–92

C

Campaigns, 85–87, 171
Contact phase of the accident
    contact phase control, 51
    energy transfers, 52
    trends, 52
Creative bookkeeping, 78

D

DART rate, 66–67
Direct losses
    compensation, 17
    medical, 17
    permanent disability, 17
    rehabilitation, 17–18
Disabling injury severity rates
    (DISR), *69*
    average days, 69
    total disabling injury, 69
    work-related accidents, 68–69

E

Energy exchange
    accident sequence, 105
    accident types, 106
    agency, 107–108
    contact control, 108–109
    energy transfer analysis, 106, *107*
    event and consequence, 144–145, *145*
Energy release theory, 105–106

F

Fatal accident rate (FAR), 76
Fatality-free shifts, 61–62
First aid
    defined, 69–70
    injury rate, 70

G

Guidelines to record injury
    deadliest projects, 61
    employment, 59
    Golden Gate Bridge, 60
    high-rise construction fatalities, 61
    historic measure, 60
    Hoover Dam, 61
    occupational fatality rates, 60
    occupational injury, 59
    World Cup 2022 Qatar, 61

H

Hazard identification and risk assessment (HIRA) process, 10
    application of, 166
    audit of, 166
    audit protocol, 168, *169*
    controls, 165
    key performance indicator, 168
    ongoing assessments, 165
    purpose of, 164
    risk assessment, 165–166
    risk register, 166, *167*
    sources of, 164
Hazards
    biological hazards, 157
    chemical hazards, 158
    classifications, 9, 158
    definition, 9
    ergonomic hazards, 158
    identifying and measuring, 151
    physical hazards, 158
    potential hazard, 9
    profiling, 158
    psychosocial hazards, 158
    safety hazards, 157
HazCom, 182
Health and safety committees
    committee system, 171
    definition, 170
    functions of, 171
    joint health and safety committee, 170
    leadership, 171
    measurable management performance, 170
    meeting agenda, 171–172
    meeting minutes, 172
    representatives, 170–171
    vital SMS component, 172
Health and safety representatives, *176*
    appointment, 175
    duties of, 175
    positive performance indicator, 176
    safety communication principle, 174
    safety participation principle, 174
    safety recognition principle, 174–175
    training, 175
Health and safety training
    accident and incident recall, 184
    accident root causes, 181
    annual refresher training, 182
    competency-based training, 183
    coordinators, 182
    critical task training, 183
    formal first aid training, 182
    formalized, 184
    hazard communication, 182

Index

induction training, 181
initial and ongoing selection, 183
key performance indicator, 184
safety management principles, 181
technically orientated, 183–184
High-risk behavior (unsafe act), 144
   accident investigation form, 145–146
   accidental loss causation, 28, 29
   complex situation, 147
   incident recall program, 153–154
   near miss incident, 147
   safety myth, 146–147
High-risk workplace condition (unsafe condition)
   accident investigation form, 145–146
   accidental loss causation, 28, 29
   categories, 150
   incident recall program, 153–154
Human failure
   active failures, 148
   deliberate failures, 148
   errors, 148
   exceptional failure, 148
   exceptional violations, 150
   inadvertent failure, 148
   latent errors, 149
   latent failures, 148
   mistakes, 149
   procedural violations, 150
   routine violations, 149
   slips/lapses, 149
   unintentional violations, 149
   violations, 149

I

Iceberg effect, 113, *114*, 134
Imbedded culture, 81
Impressive periods, 84
Incident recall program, 153–154
Injury and illness data
   advantages, 58–59
   calculations, 66
   cost of, 76
   disadvantages, 59
   OSHA recordable injury, 66
Injury-free periods, 84
Injury frequency rates
   comparsion, 63
   history, 63
   legal rating, 64
   national figures, 64
   universally used, 63
Injury incidence rates, 65, 70
Injury manipulation, 78

Inspections
   guidelines, 159
   health and safety representatives, 160, *160*
   housekeeping competition, 161, *162*
   informal inspection, 161
   legal compliance, 160
   performance metrics, 156
   planned inspections, 161
   risk assessment, 160
   safety audit, 162
   safety department, 161
   safety inspections, 157
   safety observations, 161
   safety survey, 161
   specific equipment, 162
   workplace to workplace, 162–163
International culture, 81

L

Lagging indicators
   accidental property damage, 95
   accident outcomes, 94–95
   accident ratios, 95–102
   agency, 73–75
   awards, 89–93
   body parts injured, 75
   bridging the gap, 95
   energy exchange, 71
   energy transfer (exchange) types, 71–73
   environmental harm, 103
   Exxon Valdez oil spill, 103
   fire damage accidents, 103
   high-risk behavior, 71
   major accidents, 103
   property damage accidents, 102
   safety performance, 94–104
   safety targets, 84
   totally hidden costs, 103–104
   unintended events, 95
Leading safety key performance indicators (KPIs)
   audit results, 210–211
   employees programs, 212
   fire drills, 211–212
   near miss incident reports, 208–209
   number of safety representatives, 212
   plant (workplace) inspections, 210
   requirements, 207, *208*
   responsibility and accountability, 207
   safety committee, 207–208
   safety observations reports, 209–210
   site risk assessments, 212–213
   source sheet, 213
Likert scale, 179–180
Loss causation model, 145, *146*

Lost time injury
  acid test, 63
  definition, 62
  regular job, 62–63
  shift lost, 62
  temporary total disability, 62
Lost time injury frequency rate (LTIFR), 64, 64–65, 65

M

Management safety functions
  authority, responsibility, and accountability, 22–23
  commendation for compliance, 23
  correct deviations from standards, 23
  evaluate conformance to SMS standards and achievements, 23
  identify control and mitigate risks, 20
  identify the hazards and assess the risk, 20
  measurement against the standard, 22
  risk-based, management-led, audit-driven sms, 20–21
  safety controlling, 19
  set standards of accountability, 21
  set standards of measurement, 21
Measurement of control, 59
Measures of consequence
  after the event, 58
  consequences, 58
  outcomes, 58
  rear-view mirror, 58
Measures of failure, 57
Medical treatment, 70
Mine Safety and Health Administration (MSHA), 116
  incidence rates, 67–68
  reportable injury, 67
Minor injury incidence rate (MIIR), 70–71

N

National Occupational Safety Association (NOSA)
  award plan, 89–90
  collection of injury statistics, 89
National Safety Council (NSC) (US)
  Million Workhours Award, 90–91
  Perfect Record Award, 90
  Significant Improvement Award, 90
  superior safety performance award, 90
Near miss incident
  accident ratio, 134–135
  allocate targets, 141
  benefits, 135
  calculations, 140
  closed calls, 134
  feedback, 141
  5-point safety checklist, 138, *138*
  formal reporting system, 135–136
  hazard reporting system, 137–138
  iceberg effect, 134
  informal reporting, 138–139
  leading performance indicators, 133–134
  measurement of, 135, 139–140
  metrics, 141, *142*
  monthly totals, 140, *141*
  record log, 136, *136*
  rectification system, 140
  report form, *136*
  risk ranking, 139, *139*
  safety performance measurement, 141
  safety reporting hotline, 138
No blood–no foul, 81–82
NOSA 5-Star Safety and Health Management System
  OSHA VPP, 91–92
  Safety Effort and Experience, 91–92
  Star Grading Awards, 91
  VPP star program, 92–93

O

Occupational hygiene
  hygiene hazards and stress, 9
  objective of, 8
Occupational Safety and Health Administration (OSHA), 16, 30, 38, 63, 66, 67, 69, 70, 78, 79, 92, 94, 116, 124, 186, 200

P

Post-contact phase control
  accident-based metrics, 54
  measurement, 53–54
  system failure, 54
Pre-contact phase control
  hazard burden, 51
  minimum standard detail, 50–51
  positive performance indicators (PPIs), 47
  pre-contact control, 47–48
  scoring method, 50, *50*
  SMS elements, 48–49
Property damage accidents
  costing, 102
  number and frequency, 102

R

RIDDOR, 68, 79
Risk matrix, 10–11, *11*
The Royal Society for the Prevention of Accidents (RoSPA), 34

Index

S

Safety awards, 83, 84, 90
Safety bribery
    injury-free bonus, 82–83
    root of the problem, 83
Safety fear factor, 41, 80–81, 99
Safety gimmicks, 85
Safety incentive schemes, 84
Safety management
    accurate indicator, 3
    consequence-based measurements, 3
    false sense of security, 4
    no control, 5
    safety performance, 4
    shifting the paradigm, 5
    traditional measurement, 4
Safety management performance
    appropriate mix of measures, 37
    benefits of, 36
    characteristics of, 41
    external measurements, 38
    focus, 37
    fuzzy concepts, 39
    hard measurements, 36
    input measurements, 39
    internal measurements, 37
    lagging measurements, 35
    leading measurements, 35
    management review, 37
    measurements, 33–34
    output measurements, 40
    process measurements, 40, *40*
    qualitative indicators, 34
    quantitative indicators, 34
    risk profile, 39
    safety as a state, 38
    safety effort and experience (SEE), 36
    safety record, 41–42
    soft measures, 35–36
    too simplistic, 38–39
Safety management system (SMS)
    accidental loss causation, 32
    accident investigation, 128–129
    accountability, 125
    authority, 124
    checklist, 159
    commendation, 129
    continuous process, 18
    control and mitigate risks, 122
    correct deviations from standards, 128
    critical task observation, 126
    definition, 121
    external audits, 127
    goal, 86
    goals and objectives, 86
    HIRA process, 122
    identification, 121
    indicators, 121
    internal audits, 127, *128*
    key performance indicators (KPIs), 156
    management control function, 121
    management practice, 129
    measurable performance standards, 123–124
    measuring against standards, 159
    measuring performance, 126
    objective, 86
    observation program, 151–153
    proactive objectives, 129
    responsibility, 124–125
    scoring system, *129*
    standards of accountability, 124
    strategy defined, 86
    system standards, 126
    weighting systems, 127–128
Safety management system (SMS) audit
    auditable organization, 201–202
    auditor's experience, 205
    auditor's training, 204, 205
    audit protocol, 203, *204*
    baseline audit, 199–200
    benefits, 198–199
    centralized and decentralized, 201
    compliance audit, 201
    compliment, 206
    description, 197
    documentation and records, 200
    guidelines for auditors, 205
    independent audit, 200
    internal audits, 200
    lead auditor, 205
    learning opportunity, 199
    proactive approach, 198
    questioning techniques, 205
    reactive *vs.* proactive measurement, 198
    requirements, 202–203
    safety inspection, 197–198
    subjective *vs.* objective, 198
Safety observation program
    appointed observer(s), 151
    checklist, 151
    discussions, 153
    employees observation, 152
    observation metrics, 153
    observations process, 151, 152, *152*
    record findings, 153
    report, 153
    safety observation card, 152
    time of observation, 151
    tracking system, 153

Safety perception survey
  areas, 178–179
  feedback, 180
  good perception survey, 178
  hazard identification, 177
  improve health and safety, 177
  leading indicator, 178
  Likert scale, 179–180
  purpose of, 177–178
  reevaluation, 180
Safety performance measurements
  fatal and severe injuries, 76
  road safety, 75
Safety publicity boards, 85–86
Safety record
  differing rules, 41
  public perception, 41
  safety fear factor, 41
  safety paradigm, 42
  safety *vs.* injury, 42
  world safety record, 42

T

Target injury frequency rate (TIFR), 37, 66, 71
Task risk assessments, 187–188, *188*
Toolbox talks
  attendance sheet, 187
  meaningful, 187
  OSHA, 186
  structured, 187
  topics, 186
Total case incidence rate (TCIR), 67
Total recordable disease frequency rate (TRDFR), 66
Tye–Pearson theory, 96–97, *97*

U

Underreporting of injuries
  GAO report, 79
  hidden tragedy, 79

W

Workplace health and safety
  accident, 10–11, 13
  accident causation theories, 14
  accident sequence, 14
  audit-driven, 19
  business interruption, 11
  death statistics and injury, 7–8
  direct losses, 17–18
  hazard, 9
  health and safety management system (SMS), 18
  hierarchy of control, 11
  high-risk behavior (unsafe act), 14
  HIRA process, 10
  incident confusion, 13
  injury, 13, 14
  loss causation sequence, 15–16
  losses, 16
  management-led, 18–19
  management safety functions, 19–23
  near miss incidents, 12–13
  occupational hygiene, 8–9
  risk assessment, 10–11
  risk-based, 18
  risks, 9
  safety, 8

Z

Zero harm, 86, 87